Tatting Shuttles of American Collectors

Tatting Shuttles

of American Collectors

Heidi Nakayama

First Edition

Published by Envirocare
5459 E. Atchison
Fresno, CA 93727

Internet ordering & information:
Http://home.att.net/~nakayamas/
Email: Tatkrazy@att.net

Printed by:
DeHART's Printing Services
3265 Scott Blvd.
Santa Clara, CA 95054
www.deharts.com

ISBN 0-9720628-0-7
Library of Congress Control Number: 2002091594

Dedicated to my wonderful Aunt Stella

Contents

Foreword

This project was born from my love for tatting and began with a gift one Christmas from my tatting instructor and very dear Aunt Stella Estey. The gift was a large celluloid tatting shuttle which ignited the question, "When was this made?" From this modest start began a quest that has taken me to flea markets and antique stores around the country to find more of these precious old shuttles and to discover when they were made.

After a couple years it became clear that this would be no small undertaking. When attempting to date tatting shuttles I learned that much of the historical information could not be found. In order to date many of the unmarked shuttles one must first identify the materials from which they were made. This challenged me to become familiar with many areas of U.S. history and manufacturing. For years I have compared working on this book to a gigantic jig saw puzzle because the pieces of information are often small and hard to find.

There is still much to learn in this fascinating field and perhaps with time additional information will be found for a future edition.

Acknowledgments

As with projects of this type, there are a great many people to thank. My thanks begin with the first group of tatters I met on the internet in 1990. This group of tatting friends, from all across the continent, encouraged me to follow a dream and preserve a small part of tatting history.

My debt of gratitude includes many patient librarians who took the time to answer my inquiries, often going beyond the call of duty to help me find what must have seemed liked obscure information to them.

My thanks also go to the tatters and sewing tool collectors here and abroad who sent me endless photocopies of shuttle advertisements and articles.

Special thanks go to the Museum of American History, Archives Department in Washington, D.C. for allowing me access to the Warsaw collection of trade catalogues, to Sandra Ghosh, my second set of eyes while I was there, and to Sue Jessen, who opened her home and hospitality to me in this beautiful city.

My greatest debt of gratitude is to the collectors who allowed me to photograph their collections, particularly Ms. Tollinger and Ms. Woodyard, and to those who sent me pictures to use in this book. I am especially grateful to Patricia Newman for her friendship, her invaluable proofreading, and for her help in finding at least half of the shuttles in my own collection.

Last, but certainly not least, my sincerest appreciation and gratitude go to my friend Ms. Middleton, who allowed me to use photographs of her collection, and also helped me through the editing and learning processes of producing a book.

Introduction

Knotting is generally considered to be the precursor to tatting. It was essentially the vogue needlework around the middle of the eighteenth century. Knotting shuttles showed off graceful hands and were considered acceptable gifts for suitors to give young ladies. Ranging from simple and plain to very ornate, shuttles have been found decorated with intricate inlay work on ivory, mother-of-pearl and tortoiseshell. Some shuttles were made in filigrees of silver or gold, and some were carved of steel which was then gilded. Knotting shuttles were not known to be made in America, but they are found in many American collections. Knotting shuttles vary from four inches to six inches in length and have a large space between the sides used for the winding of heavy silk and linen thread. Examples are shown on color plate 1. For more information on European knotting and tatting shuttles, watch for an upcoming book by Pam Palmer.

There is much speculation about where and when tatting first began. Although patterns for tatted lace first appeared in the 1850s, it is generally believed that tatting started much earlier. However, the majority of shuttles shown here were made in the late nineteenth and twentieth centuries.

This book will focus on the enormous variety of shuttles that are found in American collections today.

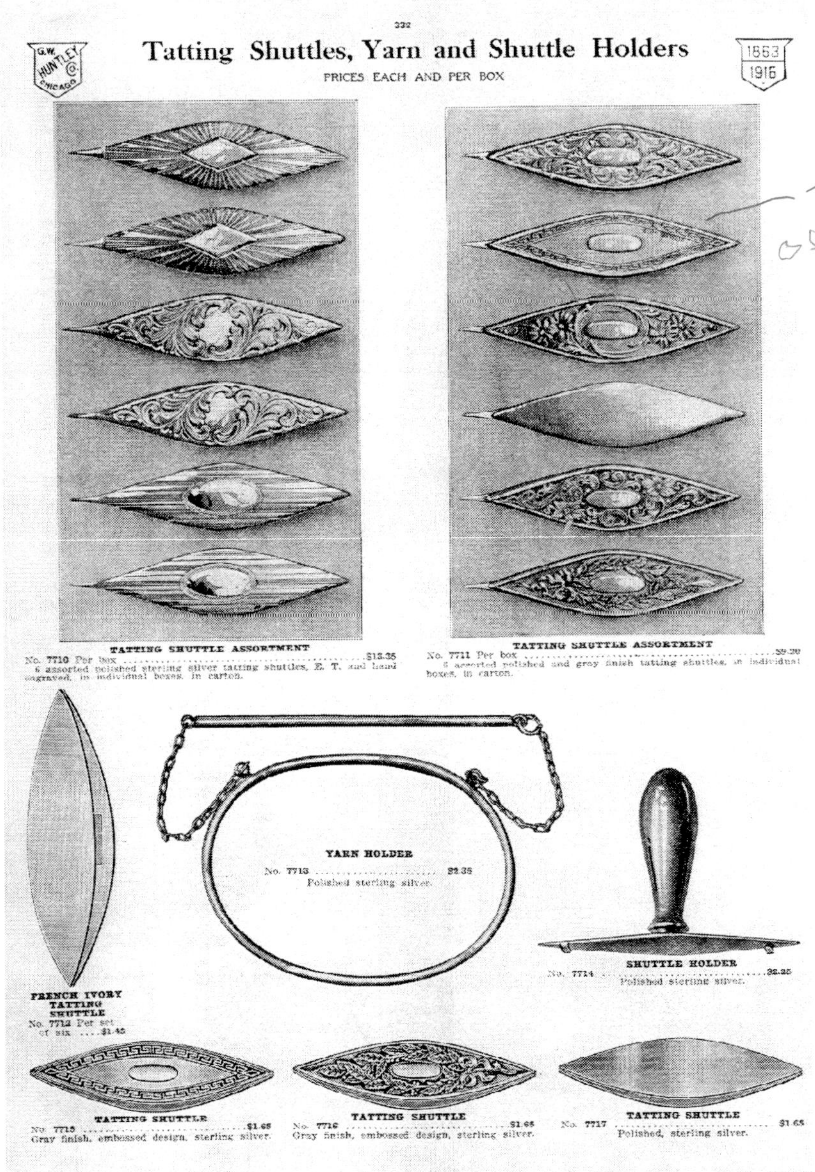

1-1 1916 G. W. Huntley Company of Chicago catalogue

Chapter One

Sterling Silver, Gold and Enamel Shuttles

The sterling silver shuttle is the epitome of American shuttles. It was the best America had to offer when tatting was at its highest popularity from 1910 to 1940. Until the U.S. Stamping Act of 1906 there was no federal regulatory system for the purity of silver. Instead, the silver industry set its own standard, much like the guilds in England. After 1906, for an object to be marked with the word "Sterling," it had to have a purity of 925 parts of silver per one thousand, the remaining 75 parts could be any alloy.

Another mark usually found on sterling shuttles is the hallmark. A hallmark is the signature or logo of the silversmith. Hallmarks were usually stamped along the outside edge of the shuttle toward the tip; however, they can be found occasionally on the inside edge, the center post or camouflaged into the design.

One of the leading retailers of silver shuttles was Daniel Low & Co., of Salem, Massachusetts. In 1867 Daniel Low opened a small jewelry store that quickly became known for its unusual gold and silver novelties and gifts. In 1893 he published his first small catalog and eventually the annual Daniel Low Year Book made the company a nationally known mail order institution. During the first few decades of the twentieth century, Low advertised extensively in magazines such as *Modern Priscilla* and *Needlecraft*. The earliest Daniel Low sterling shuttle, shown in *fig 1-1*, sold in 1898 for $1.25 and continued to be sold up until 1915. Shuttles appeared in the Daniel Low catalogues up to 1929.

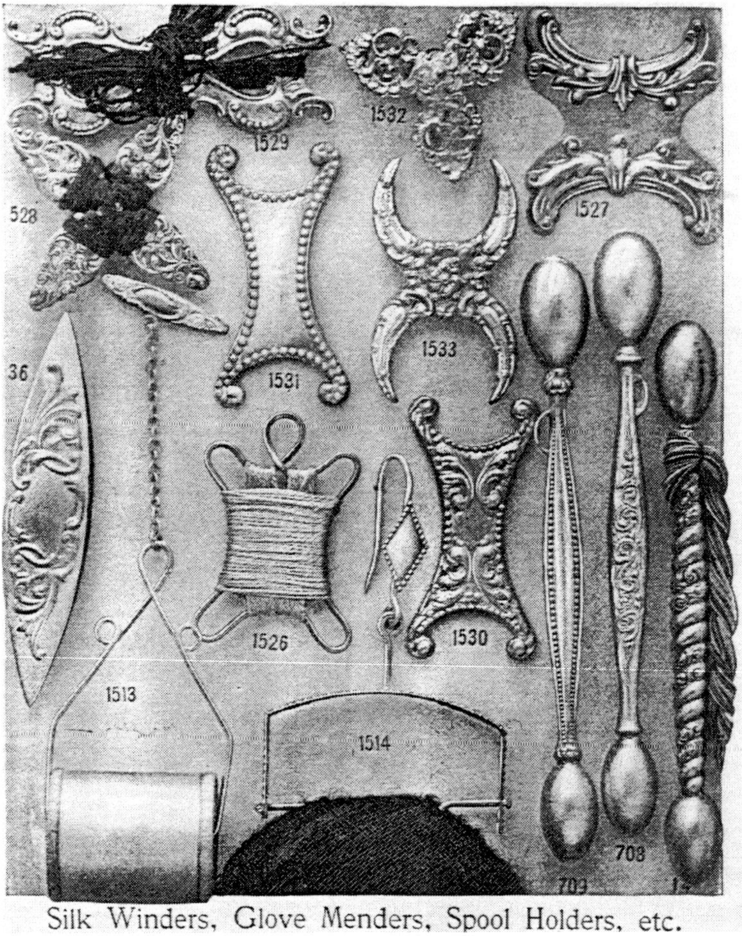

1-2 Daniel Low's first advertised shuttle 1898. Courtesy of the Peabody-Essex Museum of Salem, Massachusetts.

Gold shuttles are quite rare in comparison to sterling shuttles. They were made in several designs and were mostly hand engraved. Companies that produced these special shuttles include: Eisenstadt Mfg. Co., Leach & Miller, and an unknown company with the hallmark of "ALLCO."

Enameled shuttles began to appear around 1915. Most were enameled with a technique known as Basse-Taille. This technique begins with a stamped or engraved design in low relief on an already formed sterling plate which is then filled with a transparent colored enamel. Another technique used was Guilloche, french for "engine-turning." Engine turning is the mechanical cutting of lines on metal to create a design. Both of these techniques enables the light to reflect off the metal surface giving it a brilliant three dimensional quality.

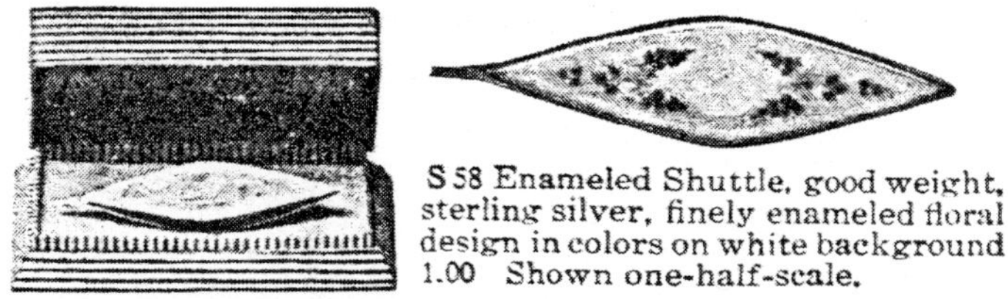
S 58 Enameled Shuttle, good weight, sterling silver, finely enameled floral design in colors on white background 1.00 Shown one-half-scale.

S 482 Enameled Shuttle, good weight, sterling silver, same size as S 57 shown above, in dainty gift box .90 Enameled in dainty plain colors: red, green, pink and lavender.

1-3 Enameled shuttles, 1925 Daniel Low catalogue.

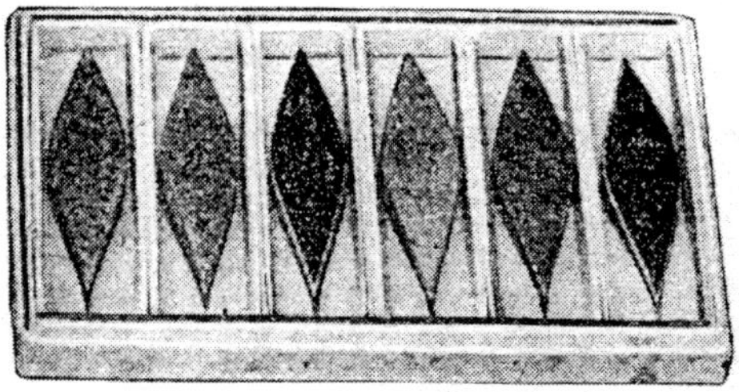
J7305 **Dozen 4.00**

Tatting shuttle; sterling silver; Cloisonne design colored enamel inlay top; a wonderful seller at .50 each; put up 1 in a box, ½ dozen boxes to a carton.

1-4 Enameled shuttles, 1916 Gray & Dudley Co. catalogue.

Silversmiths

Bob-O-Link / Eisenstadt Mfg. Co.

The sides of the Bob-O-Link shuttles were designed to pull apart to facilitate thread winding. Hidden on the inside plate of the shuttle's center post are the words; Sterling, Bob-O-Link (occasionally the word Bob-O-Link is missing), Pat. June 22, '15, and the trademark of the Eisenstadt Mfg. Co. The shuttles were patented by G.A. Donaldson, but manufactured and sold by the Eisenstadt Mfg. Co. of St. Louis, Missouri. This company also manufactured jewelry and often advertised in ladies' magazines. The Bob-O-Link shuttle could be ordered plain or with a stamped Greek key border design, in two sizes (2 1/4 or 2 5/8 inches), with or without a point, in a choice of nickel plated metal, sterling, or gold. The nickel plated shuttles were marked "Ger. Silver" and the gold shuttles were marked either 10K or 14K. Occasionally the shuttles have been found with an "engine turned" design. Most of Bob-O-Link shuttles were also crudely engraved on the opposite side of the hallmark plate with the year made in Roman numerals.

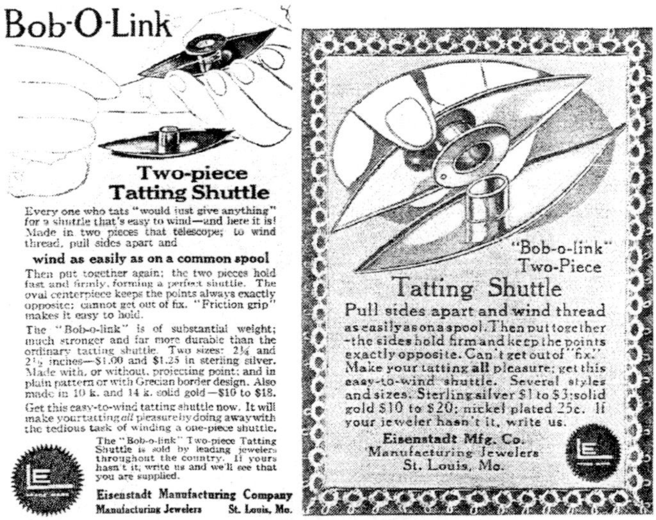

1-5 Left: 1915 Bob-O-Link advertisement, Right: 1916 ad.

1-6 Inside plate of a Bob-O-Link shuttle showing Eisenstadt Mfg. Co. trademark.

1-7 Bob-O-Link shuttles with Greek key design, Tollinger & Woodyard collections.

R. Blackington & Co.

The R. Blackington & Co. of North Attleboro, Massachusetts was started by Roswell Blackington and Walter Ballou in 1862. R. Blackington & Co. produced hand engraved sterling and enameled shuttles. Examples of the company's enameled shuttles can be seen in color plates 2 & 3.

1-8 R. Blackington & Co. shuttle, Tollinger collection.

Codding Bros. & Heilborn

C. B. & H.

STERLING.

Codding Bros. & Heilborn was founded in 1879 in North Attleboro, Massachusetts. The company was incorporated in 1893 for the manufacture and sale of jewelry, silverware and novelties. They made a variety of stamped and hand engraved designs, as well as a few enameled shuttles. The corporation went out of business in 1918.

1-9 C.B.& H. sterling shuttle, Demmer collection.

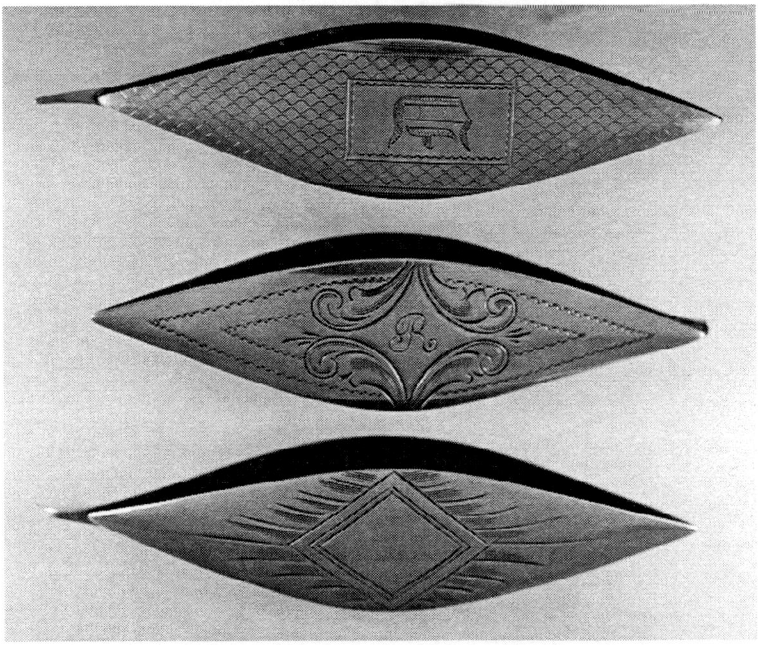

1-10 Codding Bros. & Heilborn shuttles, Tollinger collection.

Foster & Bailey

Foster & Bailey was established in 1878 after Walter E. White withdrew from the White & Foster Company of Providence, Rhode Island. They continued to operate as the Theodore W. Foster & Bros. Co. from 1898 until 1951. The tatting shuttles manufactured by Foster & Bailey were made of a much heavier sheet of sterling than most shuttles. The stamped designs all have an oval center cartouche and range from a simple bordered edge to more elaborate designs.

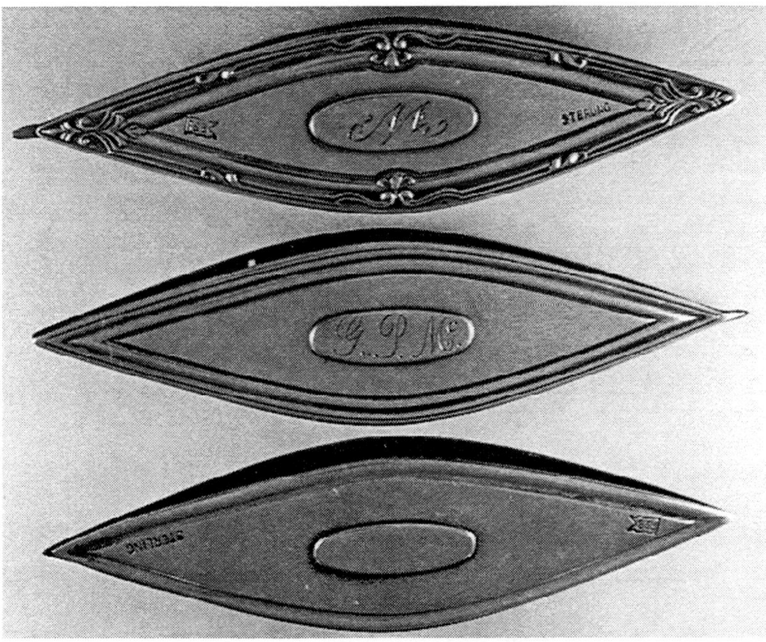

1-11 Foster & Bailey shuttles, Tollinger collection.

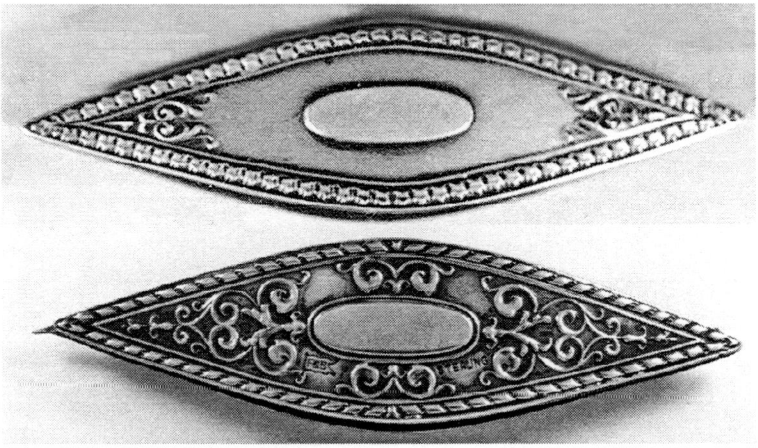

1-12 Foster & Bailey shuttles, Tollinger collection.

Hallmark

HALLMARK

The first record of the Hallmark Silversmiths, Inc. of New York City, New York, is an advertisement found in a 1917 edition of the *Saturday Evening Post.* The company merged in 1954 to become the Hunt-Hallmark Co. Pictured here is an early hammered design shuttle.

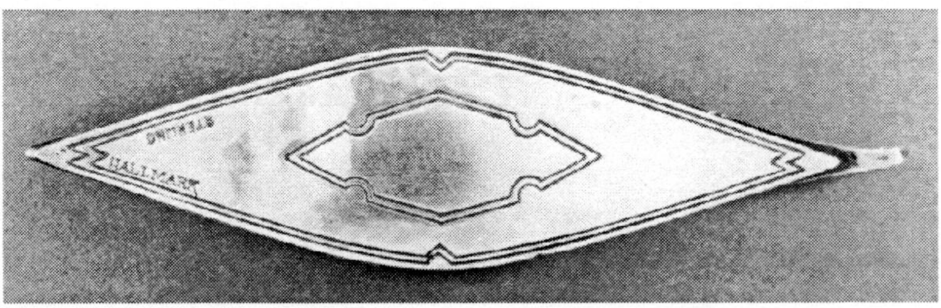

1-13 Hallmark shuttle, author's collection.

Herbst & Wassall

Herbst & Wassall of Newark, New Jersey, was listed as a manufacturer of sterling silver goods, 14K gold ware and novelties from 1904-1931. No shuttle is shown, but examples are known to exist.

Nussbaum & Hunold

N&H

Nussbaum & Hunold was listed as a jewelry manufacturer in 1920. They made a large number of shuttles, including one that was widely advertised as a premium in *Needlecraft* magazine. The company produced two sizes of shuttles with elaborate and geometric stamped designs, as well as enameled shuttles. See color plate 3.

Silver Tatting-Shuttle

Given for Six Subscriptions

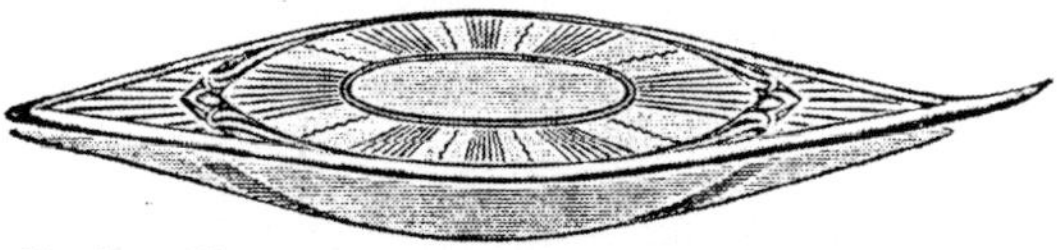

No. 6302. Sterling Silver carries its own guarantee of quality and needs no elaboration. We therefore more particularly call attention to the fact that this shuttle is fitted, in size, to the hand, has wide space for thread with close points and can be wound without a holder. Notice the pointed end for picking up picots. Illustration is full-size. Finished in dull gray.

1-14 1917 *Needlecraft* magazine premium advertisement.

1-15 N & H shuttle, author's collection.

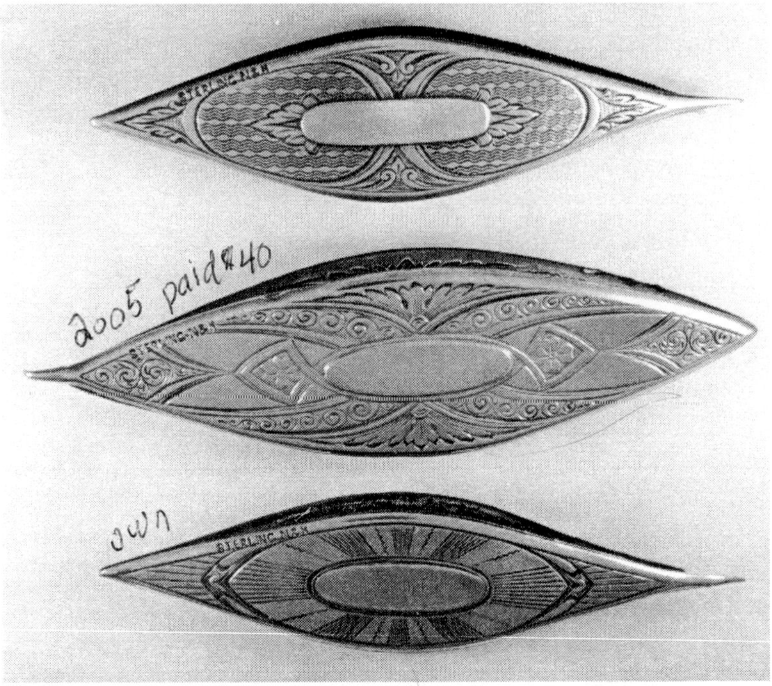

1-16 N & H shuttles, Tollinger and author's collections.

Leach & Miller

Leach & Miller Company of Attleboro, Massachusetts, produced hand engraved gold shuttles. Little information could be found on this company. See color plate 4.

1-17 Leach & Miller shuttle in gold, Tollinger collection.

S.M. Lewis & Co.

**S. M. L. & CO.
STERLING**

S. M. Lewis & Co. (aka Lewis Bros.) of New York was listed as early as 1896 and went out of business sometime before 1915. Examples of these shuttles exist but are not shown here.

Simons Bros. & Co.

The Simons Bros. & Co. was established by George W. Simons in 1840 and continues under the same name today. One of Simons' shuttles had a very unusual twin fish design which was sold by Otto Stumph & Co. in 1911.

1-18 1911 Otto Stumph & Co. Catalogue.

Thomae Co.

The Thomae Co. of Attleboro, Massachusetts was started in 1920 by Charles Thomae and his son Herbert. They made a variety of novelty items in 14K gold and sterling silver. Thomae shuttles had very distinctive hand engraved designs.

1-19 Thomae shuttles, Tollinger collection.

Unger Bros.

The Unger brothers began the manufacture of silver jewelry in 1878. Over the years the company's products ranged from flatware to novelties. The firm stopped making silver articles in 1914 and went out of business in 1919.

1-20 Unger Bros. shuttle, Woodyard collection.

Webster

George K. Webster established the G. K. Webster & Company in North Attleboro in 1869. After his death in 1894 it was called the Webster Company. In 1950 it became a subsidiary of the Reed & Barton Company of Tauton, Massachusetts but continued to operate under the Webster Company name.

The Webster Company of North Attleboro, Massachusetts, was the leading manufacturer of sterling, gold and enameled shuttles. The company sold many of their products to large retail and mail order companies of the time. Their shuttles included plain, engine turned, and elaborate floral designs. Most of these designs were stamped, but some hand engraving and hammering techniques were used as well. Usually the designs were only on one side with the opposite side plain, but occasionally both sides were decorated. The shuttles were all the same size and came with or without a point The enameled shuttles were made in a number of solid colors: blue, pink yellow, lavender and green. Webster also made enameled shuttles in a variety of floral designs. See color plates 2 & 3.

1-21 Webster Co., Tollinger collection.

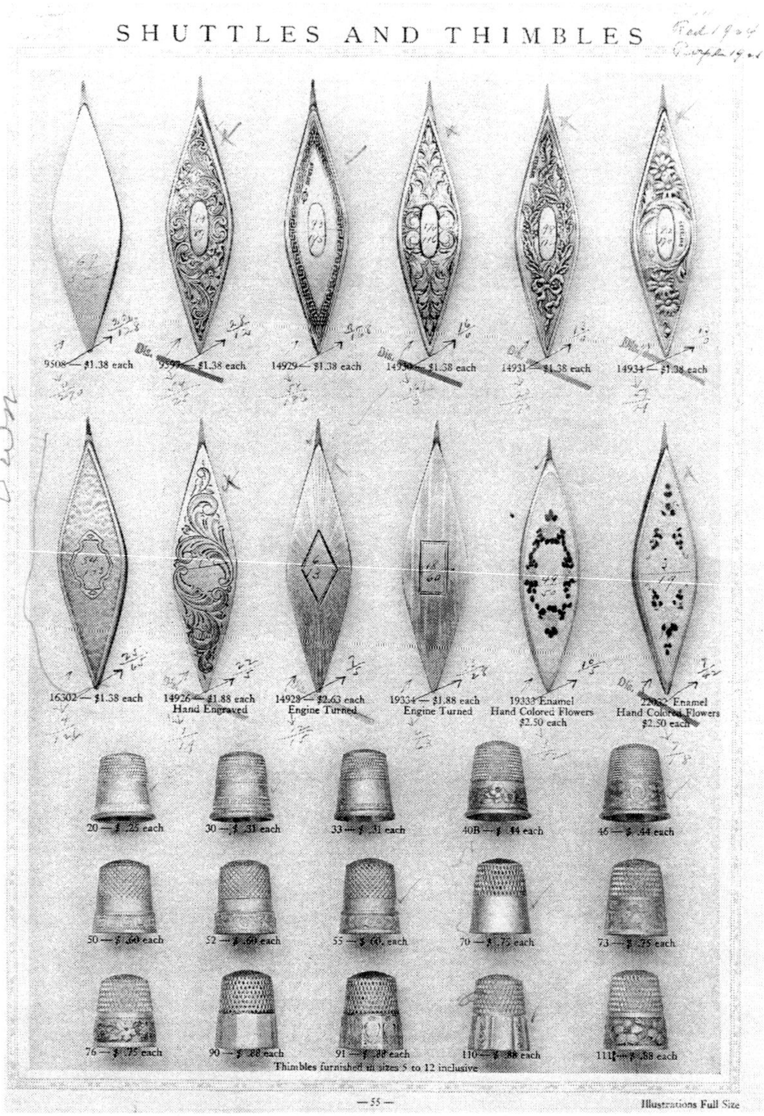

1-22 Webster Company Sterling Silverware and Novelties catalogue 1924-25. Compliments of John Kuenzig of Topsfield, Massachusetts. Kuenzig Books and Instruments, P.O. Box 452, Topsfield, MA 01983.

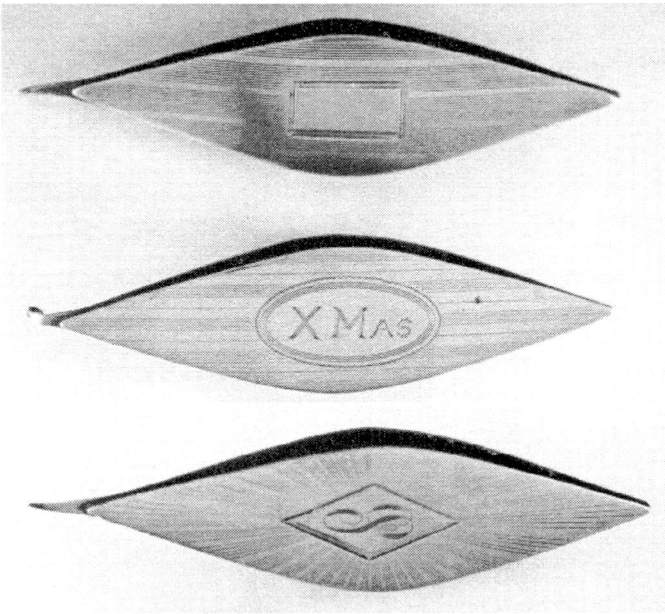

1-23 Webster engine turned shuttles, Tollinger collection.

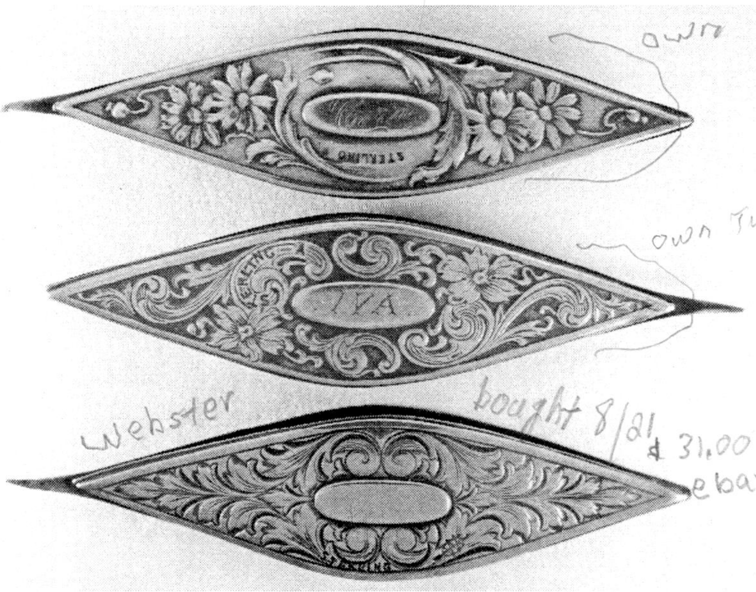

1-24 Webster sterling silver shuttles, author's collection.

Contemporary Silversmiths

Turner Scott

The winter 1955 issue of "Shuttle Art – The Magazine for Tatters" contains an article and advertisement about a Canadian silversmith named Turner Scott. He made sterling silver shuttles in both polished and hammered designs in large, medium and small sizes. It is unknown whether Mr. Scott hallmarked his shuttles. No example shown.

Eleanor Brand

ECB

From 1980 until 1994 Eleanor Brand, a metal smith and enameling teacher in Ohio, made a large variety of needlework tools in bronze, copper and sterling silver. She used a number of different techniques including chip carving, punching, piercing and etching. Ms. Brand produced 12 different tatting shuttle designs and all of her work was hallmarked with her initials "E.J.B." She was considering returning to shuttle making again, just before her death in 2000.

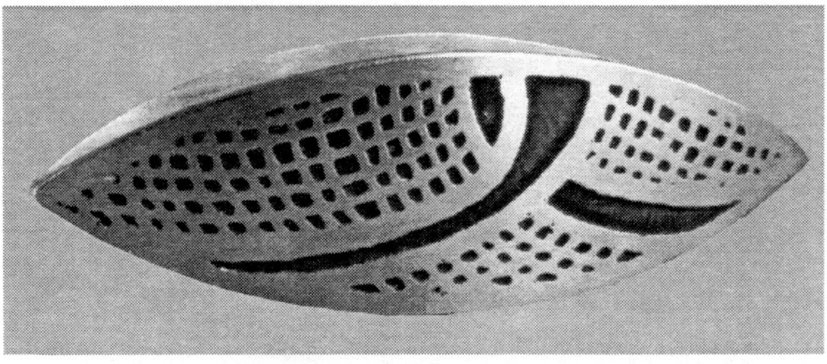

1-25 Eleanor Brand shuttle, photo by David Brand.

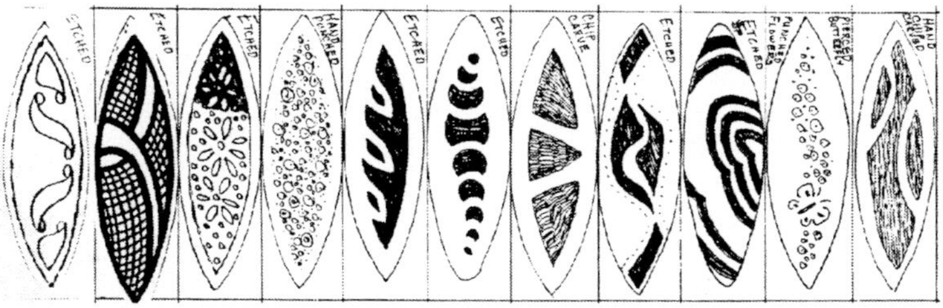

1-26 The various designs of Eleanor Brand.

Karen Bovard

KKB

The latest sterling shuttle silversmith to emerge is Karen Bovard of Nebraska. After tatting for some 25 years, and not wanting to pay the increasing prices for vintage sterling shuttles, Ms. Bovard became a self-taught silversmith of quality sterling shuttles in 1994. Her silver fabricated shuttles have a soldered post, are filed, and polished. She produces them with plain sides as well as several acid etched designs. The most recognizable is her Celtic knot design. All of Ms. Bovard's shuttles are stamped with the word "sterling" and her initials "KKB" in small block letters on the center post.

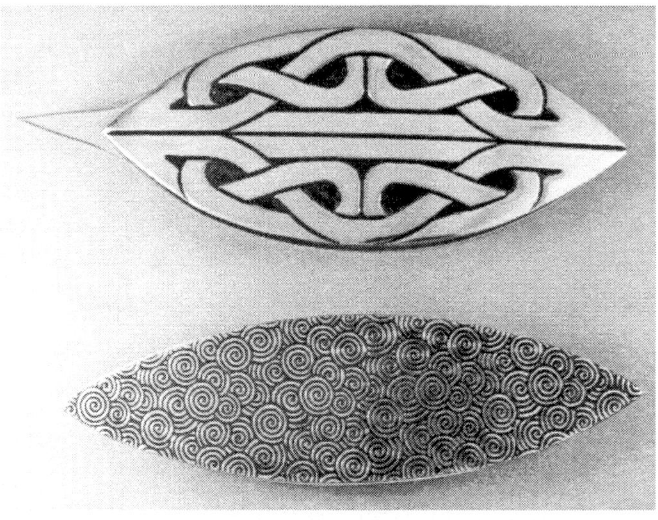

1-27 Karen Bovard shuttles. Top: maker's collection. Bottom: author's collection.

Lauren Snyder

Producing on a smaller scale is the gifted jewelry maker Lauren Snyder. Snyder is known for her tiny sterling shuttle earrings and pendants.

Unmarked shuttles

Many shuttles have been found without hallmarks. The example shown below is one of several found with a variety of native American Navajo silver markings.

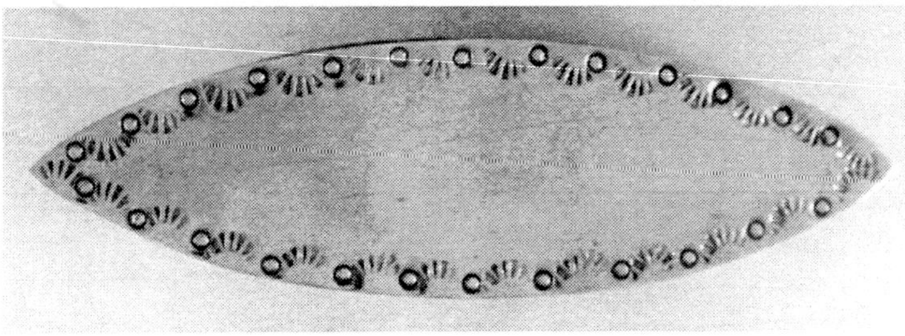

1-28 Unmarked sterling shuttle, authors collection.

Chapter Two

Nature's Best Shuttles

Ivory

The majority of ivory shuttles found in America today were made in either Europe or Asia. At the time of writing, this author has not found documentation of an American company manufacturing ivory shuttles.

The most common ivory shuttles found in America were imported from Asia. They are finely carved in Cantonese style

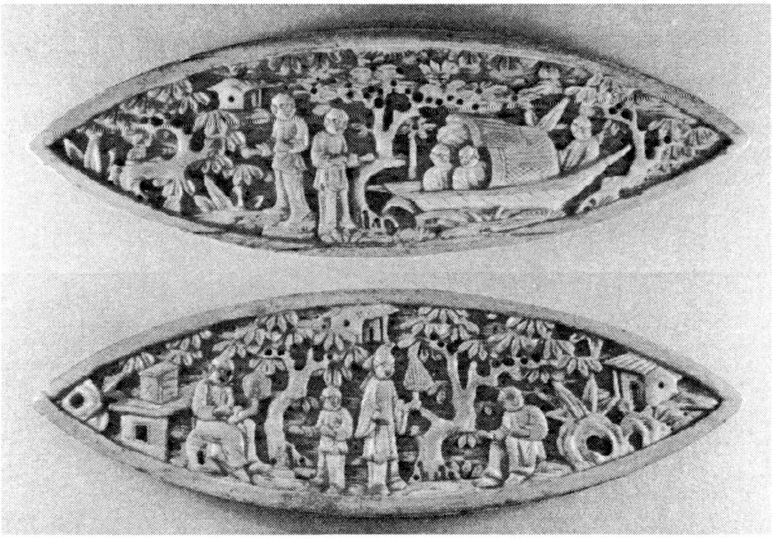

2-1 Carved ivory, 3 1/4 inches, showing both sides, Middleton collection.

with either an oriental garden scene or a four petaled lotus flower being the most common designs. These shuttles came in fancy sewing workboxes from the early to mid nineteenth century and ranged in sizes from 3 ¼ inches long by 1 ¼ inches wide to 1 ¾ inches long by ¾ of an inch wide. They were remarkably narrow on the inside, most having only 1/16 of an inch between the sides for thread. Although these shuttles are beautiful, the lack of thread space and rough exterior made them impractical for tatting. The use of ivory for any product fell out of favor around the middle of the nineteenth century, as society began to recognize the environmental impact on the dwindling elephant population.

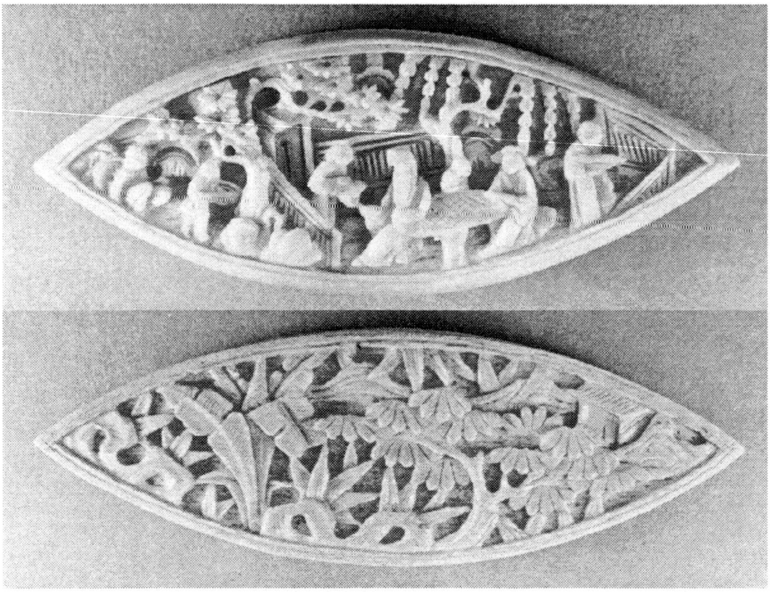

2-2 Carved ivory, Middleton collection.

In the later part of the 19th century as ivory became more scarce, the Asian carvers began substituting bone for ivory tools in the workboxes. Bone is considerably harder than ivory, so the carved bone shuttles did not have the beautiful detailing seen on the earlier ivory shuttles.

2-3 Carved bone shuttles, Tollinger and author's collection.

The only marked ivory shuttle known to this author is one imported from and stamped with the word "Denmark." These were found advertised in the fall 1958 newsletter "Shuttle Art - The Magazine for Tatters" by Dorothy Tooker and Ruth Elliott Wood of West Hartford, Connecticut. These 2 ½ inch well polished shuttles were also made in bone and could be ordered with or without a point.

2-4 Ivory shuttle made in and marked Denmark, author's collection.

Is it Ivory or Bone?

Recognizing ivory from bone can be tricky, but in most cases the difference can be seen with a magnifying glass. Both ivory and bone are porous, but in ivory the tubules that carry an oily substance are microscopic, making ivory appear solid. Bone, on the other hand, has large channels for carrying blood. This is the reason the surface of bone often appears to have dark streaks running in one direction.

Ivory is often defined as any animal tooth growing outside of the mouth. This could include hippopotamus, walrus, narwhal, sperm whale and wild boar. Some consider any tooth to be ivory. However, the purest and most widely accepted definition, is that true ivory only comes from the tusks of elephants. Tusks are the incisor teeth coming from the upper jaw of the elephant that grow from the inside in a conical manner. The layers of dentine and enamel result in curved lighter and darker parallel lines. The direction in which the tusks are cut determine the flow of these lines. Cut crosswise, these curved lines intersect themselves creating diamond-shaped spaces. Cut lengthwise, the lines may appear straight or curved, with long thin parallel clouds. These clouds are usually whiter, more opaque, and can be very wavy.

2-5 Graphic example of intersecting curved lines of ivory.

Bone is much lighter in weight than ivory. Cut crosswise bone has many minute holes, often looking like blackish dots. Cut lengthwise bone has small narrow channels running in the same direction. Depending on how a piece is finished, these channels may appear dark or light. When a piece is well cleaned and polished, the only way to see these channels is by shining light over the surface of a piece, creating a shadow that reveals the channels. Bone may also appear to have cloudy lines, but with practice one will be able to tell the difference between bone and ivory.

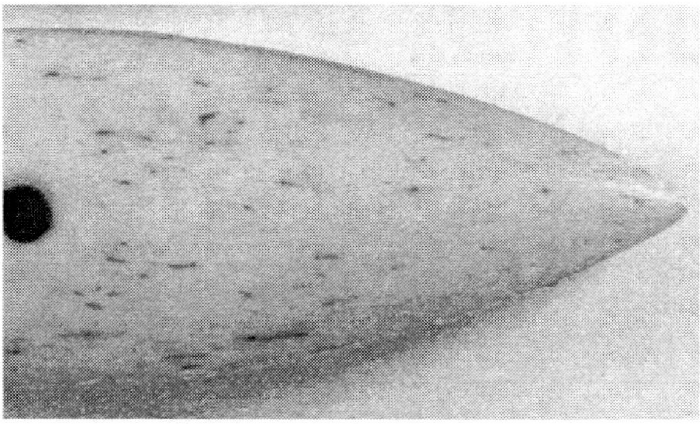

2-6 Detailed example of bone shuttle.

Bone

In America, C. J. Bates and Sons, Inc. was the only major manufacturer of bone shuttles. This extraordinary company existed for more than 100 years. Its founder, Carlton Joseph Bates, began working for the Tyler and Post Company in 1861 when he was just 14 years old. This company began manufacturing ivory artifacts in 1860 and expanded its production line in 1865 with the purchase of Griswold-Gladding Company of Essex, manufacturers of crochet hooks made from whale and cattle shin bone.

The C. J. Bates Company began in 1873 when Carlton Bates bought the business from Mr. Post. The first published proof that the C. J. Bates Co. manufactured tatting shuttles is found in the *Thomas Register* of 1905-06.

However one may find reference to Bates' 3 inch riveted bone shuttles in American catalogues and ladies' magazines as early as 1880. These shuttles are held together by two brass rivets. Most are either 2 3/4 to 3 inches in length and 5/8 to 3/4 of an inch in width. This same shuttle was also made with a variety of carved designs resembling a primitive leaf or vine. The most common of these were offered in *Needlecraft* magazine from 1916 until 1923 as a premium for obtaining one new subscriber.

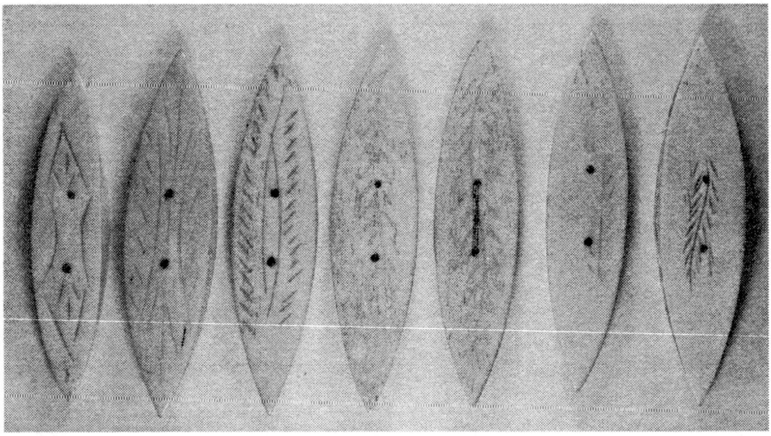

2-7 Variations of C.J. Bates carved bone shuttles used as advertising premiums. Tollinger collection.

Another popular American bone shuttle is one which was carved from a single piece of bone. The majority of these shuttles are 2 1/2 inches in length and 5/8 of an inch in width. These shuttles were sold in the years surrounding 1914.

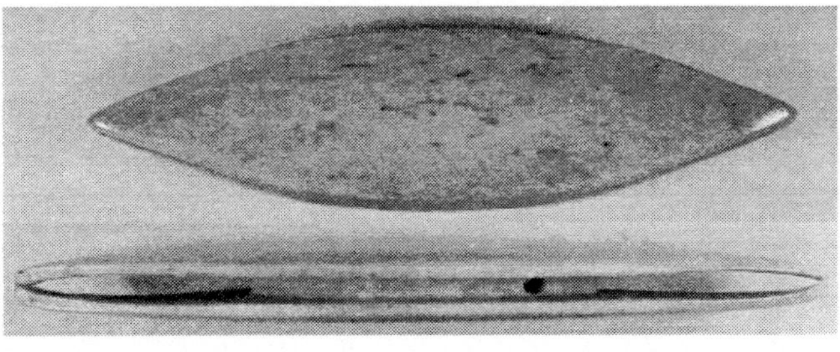

2-8 Carved one piece bone shuttle 1914, author's collection.

The 1868 E. N. Parker shuttle is the only patent known to be made from bone. It had a internal brass bobbin turned by a removable handle.

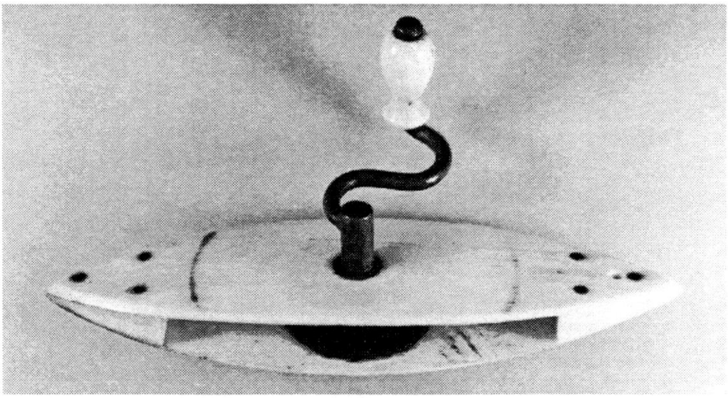

2-9 1868 E. N. Parker patent #76,512, Tollinger collection.

Carving bone and ivory shuttles was also a popular pastime for the men on whaling ships. A few have been found dated in the mid nineteenth century carved with words of love for sweethearts at home.

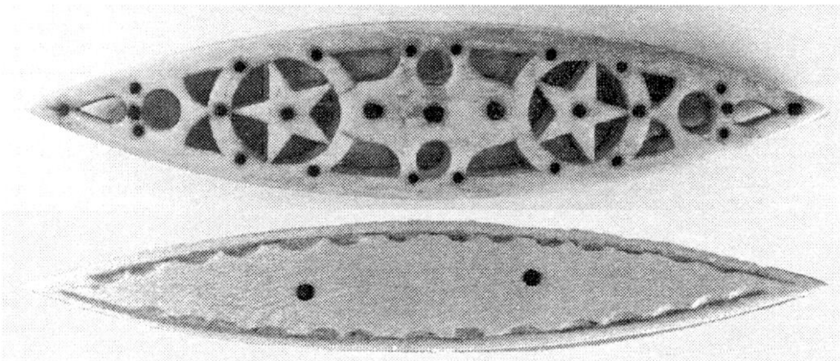

2-10 Carved ivory. Top: Tollinger collection. Bottom: Middleton collection.

Vegetable Ivory

Many small sewing accessories were made of vegetable ivory which was popular during the 1920s, however, tatting shuttles of this material are very rare. Vegetable ivory is not ivory at all but was a substitute for ivory. It is from a nut the size of a hen's egg, which comes from a small South American palm tree called Phytelephas macrocarpa. When fresh, the color and carving texture is very close to ivory. With age vegetable ivory can become golden yellow to brown depending upon how it has been stored.

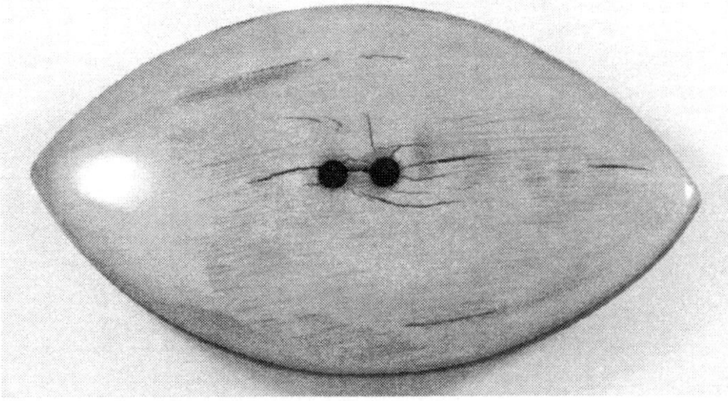

2-11 Vegetable ivory shuttle, Demmer collection.

Abalone and Mother-of-Pearl

Mother-of-pearl is a name given to the lining of the saltwater or freshwater mollusk shell. The composition is the same as a pearl. Broadly defined, it could be considered the lining of any shell, and some people also consider abalone to be mother of pearl. True mother-of-pearl is pure white, almost a blue white. However, the color of abalone ranges from the well known vivid blues and pinks to yellowish whites. Abalone also has a distinctive wavy pattern in the shell. All mother-of-pearl shuttles this author has seen are European in origin and there is no indication that an American company ever produced mother-of-pearl tatting shuttles.

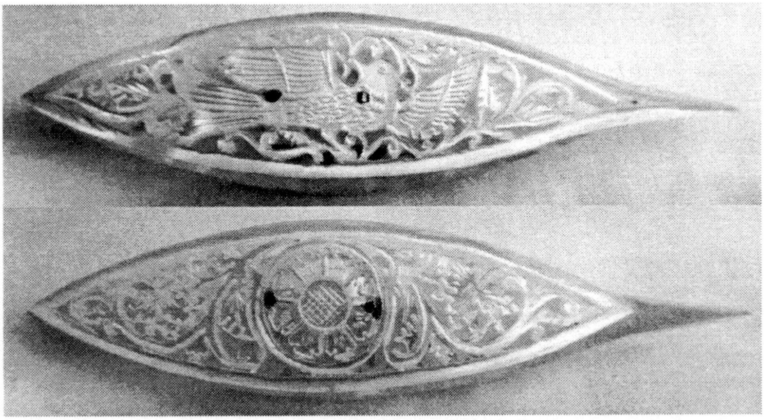

2-12 Carved and pierced mother-of-pearl tatting shuttle produced in Europe in the 1950s, Woodyard collection.

2-13 Carved mother-of-pearl tatting shuttle, Clotilda collection.

Shuttles of abalone however, were produced in America. Abalone shuttles were sold by C. C. Lord Company of Long Beach, California from 1913 to around 1925. From 1920 until 1940 they were made by Salm Mfg. of Torrance, California. Abalone shuttles were a very popular tourist souvenir for people visiting the West Coast. Many may be found still tied to the original card on which they were sold along with other sewing implements such as a stiletto, thread winder, crochet hook or tape measure. Some may still have a tourist location written on it such as "Catalina," "Long Beach," "Los Angeles" or even "Seaside, Oregon." Many shuttles were made of the whitish abalone with a smaller piece of brightly colored abalone inlaid to cover the rivets holding the center post. Occasionally abalone shuttles were made with either a carved point or a metal point that has been riveted to one side. See color plate 6.

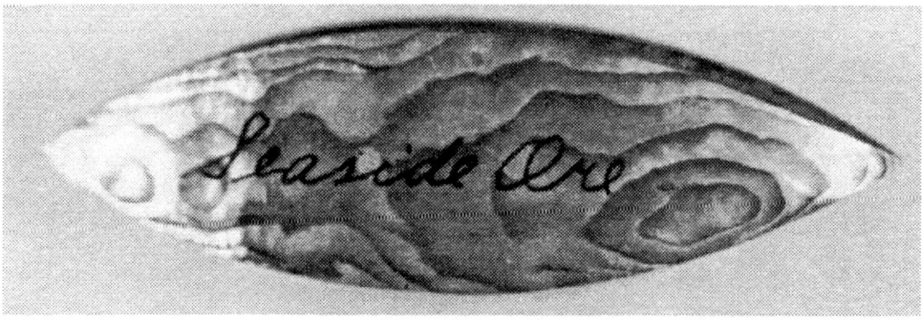

2-14 Souvenir abalone shuttle from Seaside, Oregon, Tollinger collection.

The natural oil of the skin is the best cleaner for mother-of-pearl. Rubbing one's fingers over the surface of the shuttle will remove almost any dirt or grime. Strong solvents should never be used on mother-of-pearl or abalone shuttles and they should never be soaked in water.

Tortoiseshell

True tortoiseshell shuttles were most likely made by European artisans in the 19th century. Tortoiseshell comes from the marine turtles found in the warm seas of the world with an enormous range in color and quality. Common grades of tortoiseshell are opaque in mottled colors of black and brown. More desirable is the translucent tortoiseshell that is medium brown with pale marbling. The pale transparent shell which bears a close resemblance to amber is the most desirable. There are no references of any American shuttle maker who used tortoiseshell, although marine turtles are found around Florida and the neighboring islands of the Caribbean, so it is possible that there may have been shuttles produced in America. Imitation tortoiseshell shuttles may be found made of hoof compositions and celluloid.

To clean tortoiseshell, simply wash the shuttle with soapy water and rinse thoroughly.

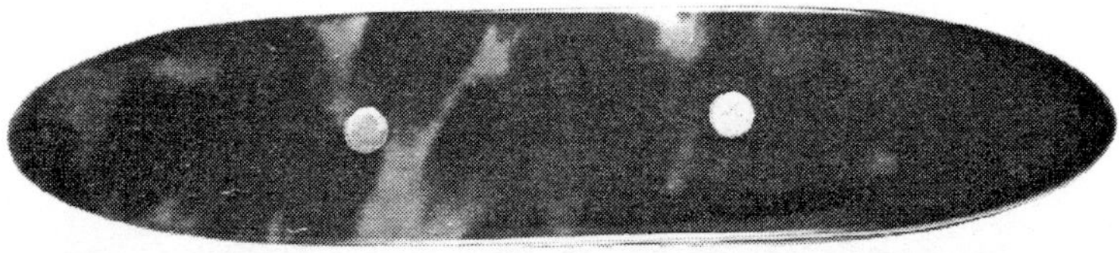

2-15 Tortoiseshell shuttle, Tollinger collection.

Horn

Both the horns and antlers of various animals have been used to make shuttles in America. These were often hand carved by folk and family artists.

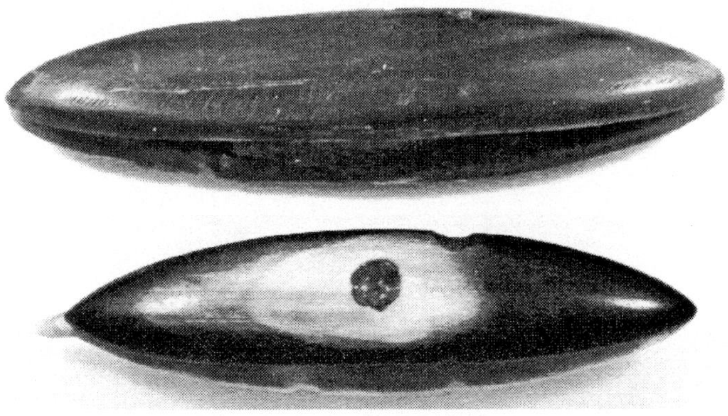

2-16 Handcrafted horn shuttles, author's collection.

2-17 European horn shuttles inlaid with abalone, Tollinger collection.

Contemporary naturals

There are a few makers today that use natural materials. Lacis, a retailer and wholesaler in Berkeley, California, imports and sells tatting shuttles made from horn, shell, bone and abalone. They are similar in size, 2 ½ inches by ¾ of an inch, and can be recognized by their slightly rounded tips and the indented edges on the center post.

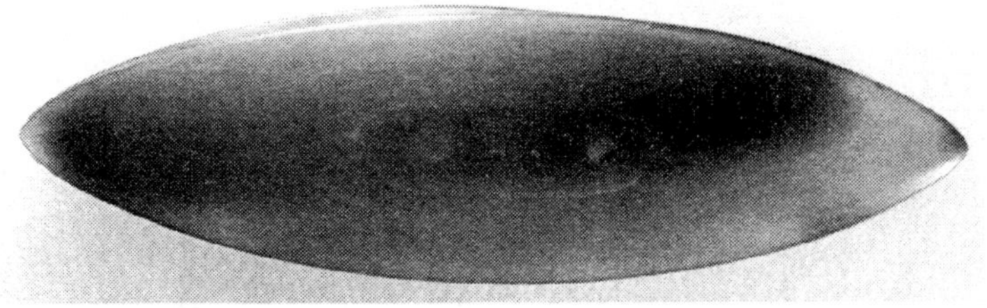

2-18 Horn shuttle from Lacis, author's collection.

Randy Houtz is a contemporary artisan in Colorado. He is best known for his carved wood shuttles, but he also carves the same unusual shuttle in horn and antler. The number "8" is found hallmarked into the side of these shuttles. See also chapter on wood shuttles.

2-19 Horn shuttle by Randy Houtz, author's collection.

Kenneth Clark of Oregon also used both mother-of-pearl and abalone in making shuttles. His shuttles have the name *Clark* and the year made on the inside of the shuttle.

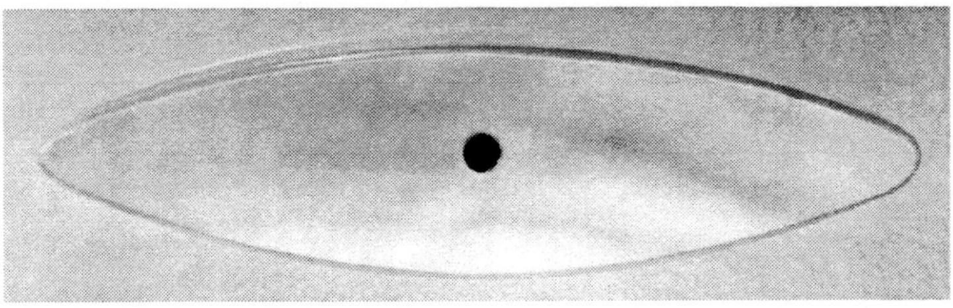

2-20 1983 Mother-of-pearl shuttle by Kenneth Clark, author's collection.

Chapter Three

U.S. Patented Shuttles

The history of tatting shuttles through the U.S. Patent office is a very interesting study. Prototypes were made for all the patented shuttles, but many of these shuttles were not feasible to manufacture in large quantities. For most tatters the simple design of the basic shuttle is all that is needed and very hard to improve upon. Despite this, there have been more than 60 patents granted for "new and useful" improvements to tatting shuttles since 1865.

All of thc shuttle patents have not been included here, however, they have been listed at the end of this chapter for those wanting to do further research. This chapter will focus on patented shuttles that have survived along with a few interesting examples. Additional information about these shuttles can be found in the chapters which focus on the material from which the shuttles were made.

The patent history of tatting shuttles begins with the Warner patent in November of 1865. It was the first of many cylinder shaped shuttles to be patented. It was also the first designed with a retractable crochet hook. This author has never found any examples of the patented cylinder shaped shuttles.

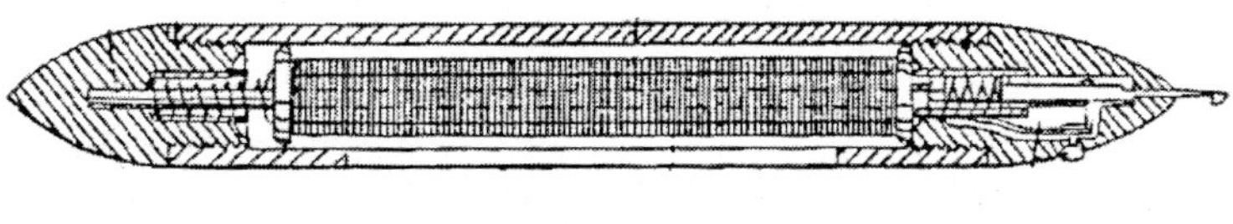

3-1 1865 J. S. Warner patent #50,970

The 1866 shuttle patent of A. Dobrowsky of Shasta, California was described as having a pin passing through it longitudinally. This shuttle was made of hard rubber and is stamped "Patent Appl. For."

3-2 1866 A. Dobrowsky patent #55,254.

The next shuttle to be made with a retractable point was the Greenwalt shuttle, patented January 1, 1867. There are several existing examples of this shuttle. The first was made of brass and decorated on one side with the patent date bordered with a floral design. The other side has a simple design over the retractable pin mechanism. It was manufactured later in hard rubber stamped with the patent date and without decorative designs, still later it was made without the patent date.

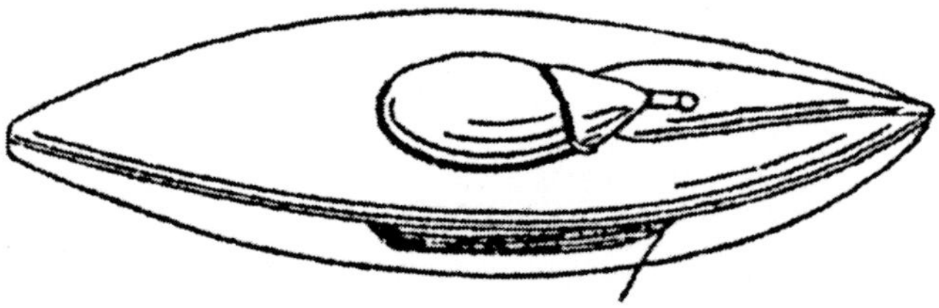

3-3 1867 A. B. Greenwalt patent #60,882.

The Parker patent of 1868 is a favorite of shuttle collectors. These shuttle were made of bone with the ends riveted together enclosing a revolving spool that was turned by a bone handled key. Unfortunately, more often than not, the shuttle is found without the winding handle.

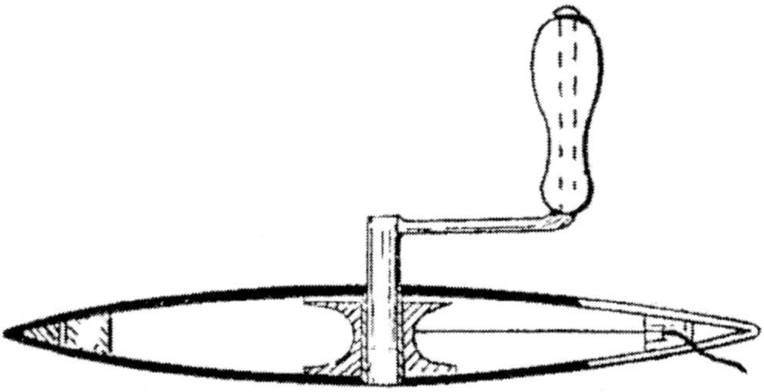

3-4 1868 E. N. Parker patent #76,512.

The next patented shuttle to be manufactured was the Stockwell & Goodwin shuttle in August of 1868. It was the first shuttle formed with a stationary point. The patent suggested that it be made of any suitable sheet metal. There are many surviving shuttles of this patent. The stamped decorative design on these shuttles may be seen in the chapter on metal shuttles.

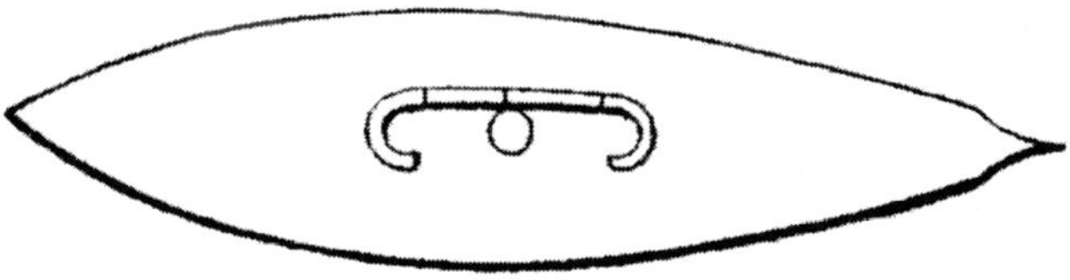

3-5 1868 Stockwell & Goodwin patent #80,781.

The Hingher shuttle and accompanying crochet hook was patented in April 1871. It was manufactured by the Novelty Rubber Co. under the Goodyear hard rubber patent. It was another closed end shuttle with a revolving bobbin, turned by crochet hook having a square shank that fit into a square socketed bobbin. The National Museum of American History in Washington, D.C., has the original Hingher shuttle and crochet hook that was submitted to the patent office. This shuttle may be seen by appointment.

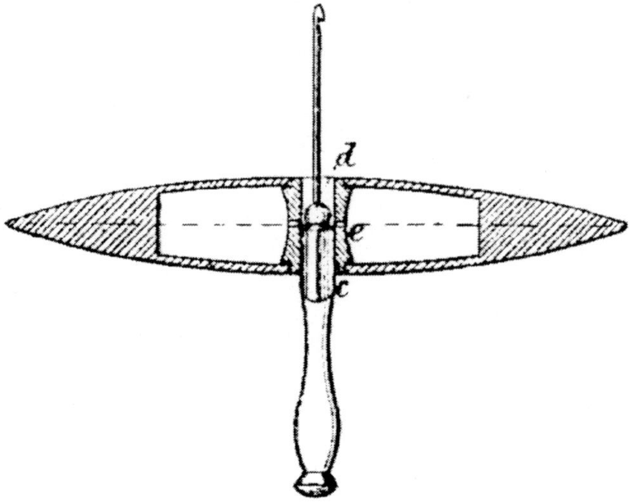

3-6 1871 Hinger patent #114,008.

Another shuttle in the collection of the National Museum of American History is the 1873 Kellogg patent. This shuttle is similar to the Greenwalt patent with the retractable crochet hook. The difference is that Mr. Kellogg's shuttle had small notches at either end of the slot for the sliding thumb knob to hold the hook extended or retracted.

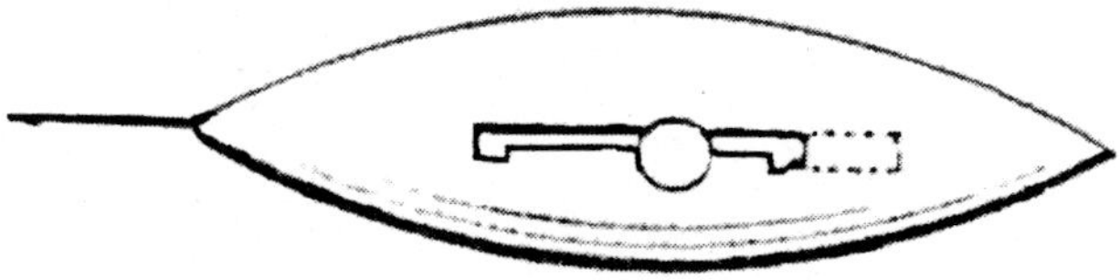

3-7 1873 E. P. Kellogg patent #146,077.

Following Kellogg's 1873 patent, there were no more shuttle patents for 41 years, until the Buff tatting needle in 1914. Because of its simplicity the Buff patented shuttle would have been easy to manufacture. However, if one were found in an old sewing basket today, most would think it was some sort of thread winder, rather than a shuttle.

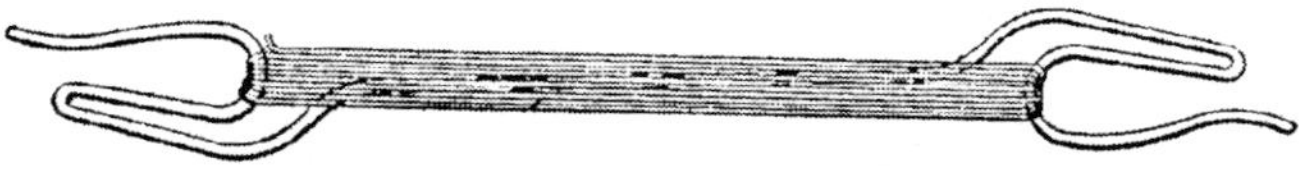

3-8 1914 E. A. Buff patent #1,116,792.

The Donaldson shuttle, patented in June 1915, was widely advertised in ladies' magazines as the Bob-O-Link tatting shuttle. It is unique in that the sides could be separated by pulling them apart. One side had an attached and fixed bobbin and the other had a sleeve that fit tightly into the opposite side. These shuttles were advertised with a wide variety of options. They could be ordered in nickel, silver or gold, two different sizes, with or without a point and also plain or with a Greek key design around the border. There were a total of twenty-four combinations. Most of these shuttles had the date of manufacture etched in roman numerals on the thread side of the spool foot.

3-9 1915 G. A. Donaldson patent #1,144,178.

Albert B. Collingbourne developed the first flat celluloid shuttle. His original design was granted a patent in 1915.

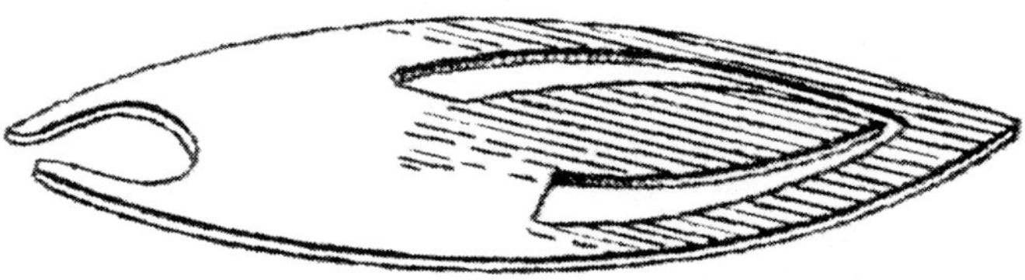

3-10 1915 A. B. Collingbourne patent #D47811

Another early celluloid shuttle to be patented was made by Charles Dill in 1916 having closed ends and a removable bobbin. Indentations on the shuttle side kept the bobbin in place.

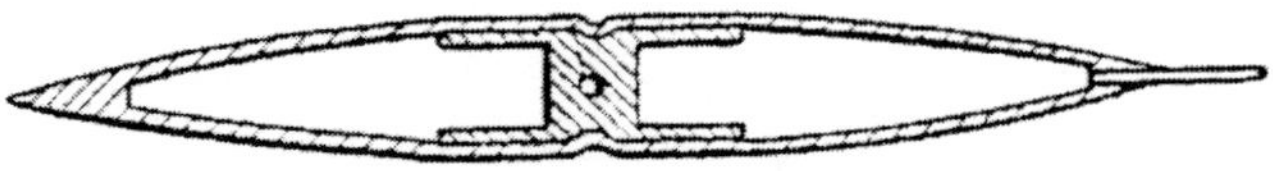

3-11 1916 C. F. Dill patent #1,178,102.

David Butler patented a popular celluloid shuttle in 1918. They were produced by the Inlaid Co. of Providence, Rhode Island. Most of these shuttles were made in French ivory, but they are also found in blue and pink.

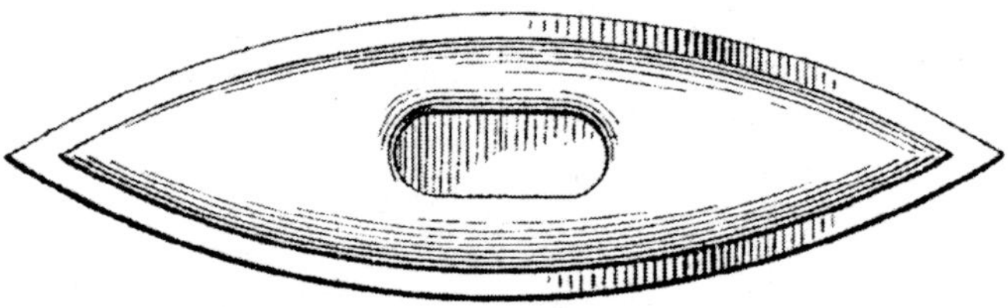

3-12 1918 D. H. Butler patent #1,272,210.

The first Justrite shuttle was patented in 1917 by A. L. Hansen who also invented and patented the "Detach-A-Bobbin" shuttle in 1918. Both of these shuttles were manufactured by the Justrite Mfg. Co. of Chicago, Illinois. The Justrite shuttle, like the Bob-O-Link, had a footed inner bobbin and a detachable side. A similar and improved patent for this shuttle was approved and given to Hamitage Richardson in 1922.

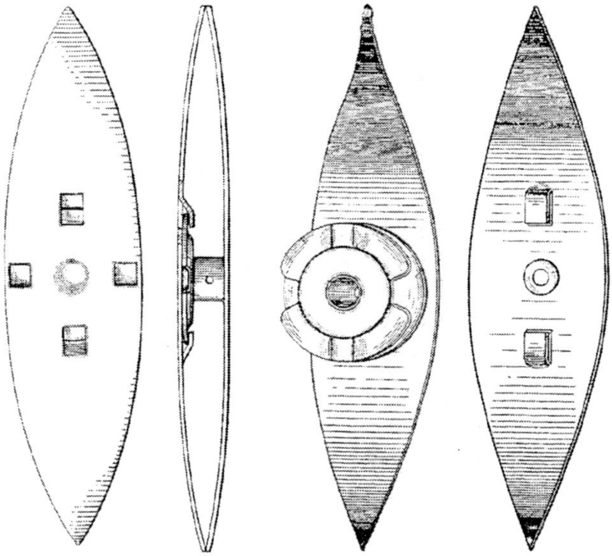

3-13 Left: 1917 A. L Hansen patent #1,215,525. Right: 1922 H. L. Richardson patent #1,453,639.

The ingenious shuttle patented in 1917 by Joseph Bigelow of Olympia, Washington was unlikely to have been produced due to its complexity. This shuttle had a spool and spring mechanism so complicated that it would wind new thread back on the spool by the creation of tension from the use and unwinding of thread during tatting.

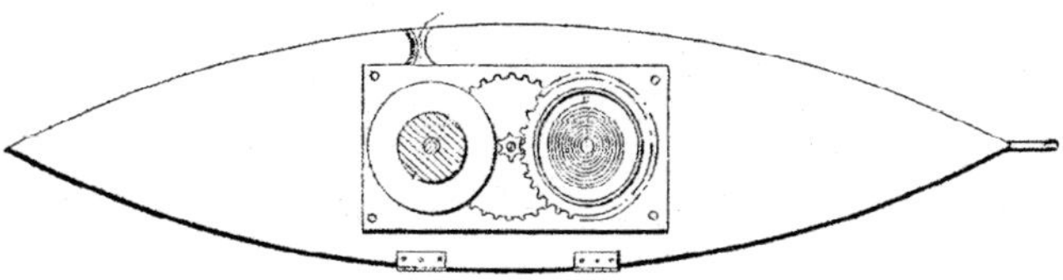

3-14 1917 J. C. Bigelow patent #1,228,286.

Undoubtedly the most recognizable name in tatting shuttles is "Boye." The first patent was given to James Boye in 1917 and was produced by the Boye Needle company of Chicago, Illinois in 1915 with the words "Pat. Appld. For" stamped on the front. The shuttles had a bobbin attached to an arm which swung out to facilitate winding or to remove the bobbin. After April 10, 1917, the shuttles are found with the patent date on it.

The Boye Needle Company's most extensively produced shuttle design was patented on April 17, 1923 by G. A. Carlson. There were two sizes for this shuttle, the larger 3 ½ inch shuttle was first sold in January of 1922. The shorter 3 inch shuttle followed in July of the same year. They were nickel plated with a hook on one end and a special projection at the other for holding the bobbin while winding. See *figure 4-6* in chapter four.

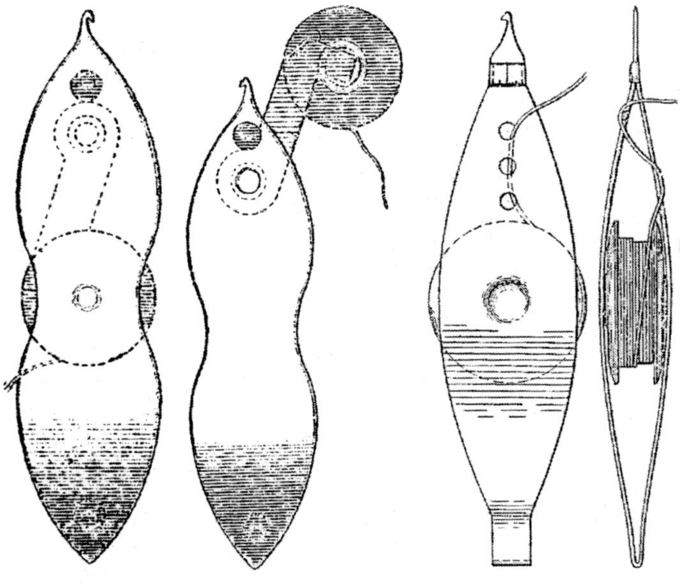

3-15 Left: 1917 J. H. Boye patent #1,222,176 Right: 1923 G. A. Carlson patent #1,452,183.

A. L. Hansen's second patent for the "Detach-A-Bobbin" shuttle was produced by the Justrite Mfg. Co. in Chicago, Illinois and also has the words "Pat. Appld. For." It is likely that this shuttle was being sold in 1915 when the patent application was filed.

This shuttle contained a removable bobbin, the long type used in early sewing machines. The machine windable bobbin was a selling point. When sewing machine manufactures changed their bobbins to their present day form, this shuttle was changed to hold the new bobbin style and the name was changed to "Detach-A-Spool."

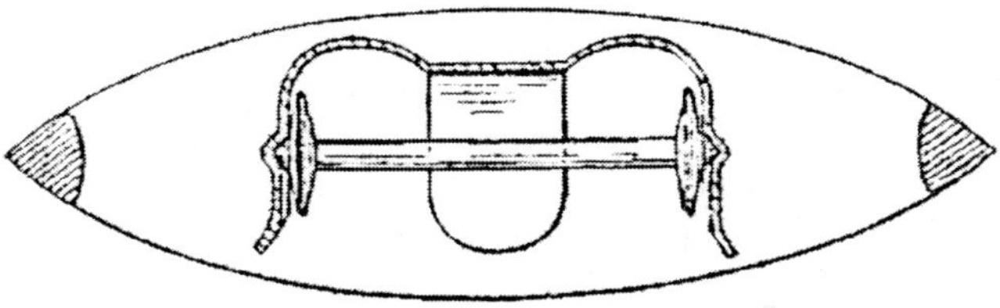

3-16 1918 A. L. Hansen patent #1,276,524.

Louis Sillman was the inventor of the first patented shuttle for C. J. Bates & Sons of Chester, Conn. in 1944. It was produced in celluloid and sold in a variety colors with the words "Susan Bates, White Dot, Pat. Applied For" stamped on the side.

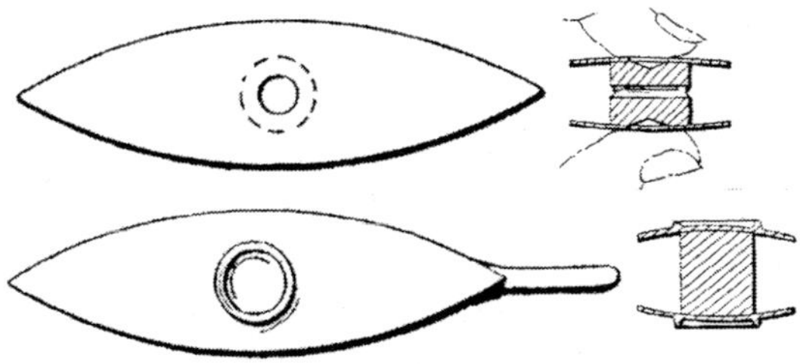

3-17 Top: 1945 L. A. Sillman patent #2,366,568. Bottom: 1944 L. A. Sillman patent #2,364,484.

G. A. Carlson's last patent for the Boye Needle Company came in May of 1947. It was an unusual flat plastic shuttle embossed with "Tatting "BOYE" Shuttle" and "The Boye Needle Company, Chicago, Pat. Pend." The shuttles were sold in four colors, green, red, white and yellow for only 6 years, from June 1947 until May of 1952.

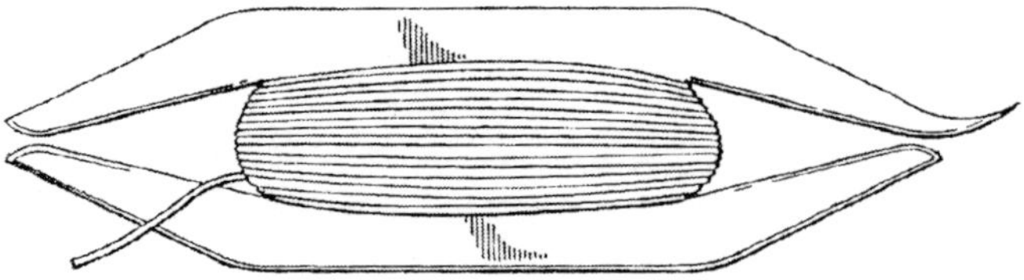

3-18 1947 G. A. Carlson patent #2,420,861.

The last patent for a shuttle of flat design was granted to Lael Morgan in 1977. It was made of a soft plastic that resembled wood. The open design accommodated the use of large thread and yarn.

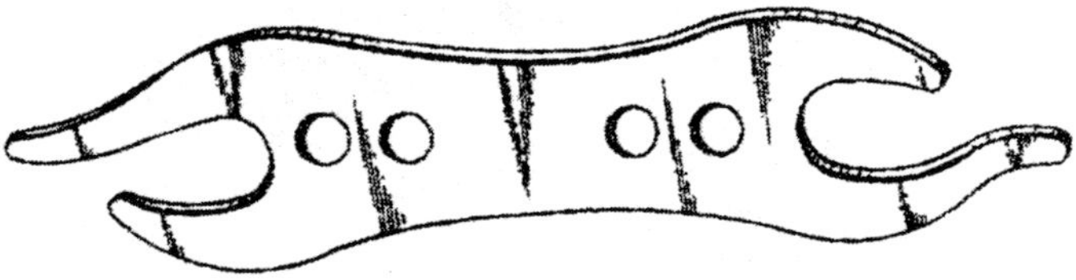

3-19 1977 Lael Morgan patent #D245,548

During the last decade of the 20th century two unique patents were produced. The Comfi-shuttle designed by Deborah Arens in 1997 is made of molded plastic. It has flanges along the outside edges of the shuttle, a grid-like grip and specially designed curve to its sides. The second patented shuttle was designed by Lauren Pelc in 1998 with two removable bobbins and is made of metal.

3-20 Left: 1997 D. Arens patent #5,642,760. Right: 1998 L. Pelc patent #5,727,439.

United States Patents of Tatting Shuttles

Copies of U.S. Patents can be obtained by writing to the Department of Commerce, Office of Patents and Trademarks, Washington D.C. 29231 or online at: http://www.uspto.gov/web/offices/ac/ido/oeip/taf/moc/jobjects/search.htm.

5,727,439	1,266,728
5,642,760	1,241,917
5,181,449	1,241,369
4,137,818	1,228,286
2,470,391	1,224,958
2,469,291	1,222,176
D245,548	1,215,525
2,420,861	1,213,214
2,395,786	1,206,555
2,373,958	1,187,306
2,366,568	1,185,534
2,364,484	1,182,681
1,632,946	1,180,227
1,551,657	1,173,102
1,521,528	1,157,535
1,473,790	1,144,178
1,453,639	1,443,238
1,452,183	1,116,792
1,443,814	146,077
1,432,184	114,008
1,411,136	106,415
1,401,053	104,565
1,390,188	103,436
1,349,741	80,781
1,345,303	76,512
1,314,209	70,273
1,306,692	69,845
1,297,234	65,309
1,287,010	60,882
1,276,524	55,254
1,272,210	50,970

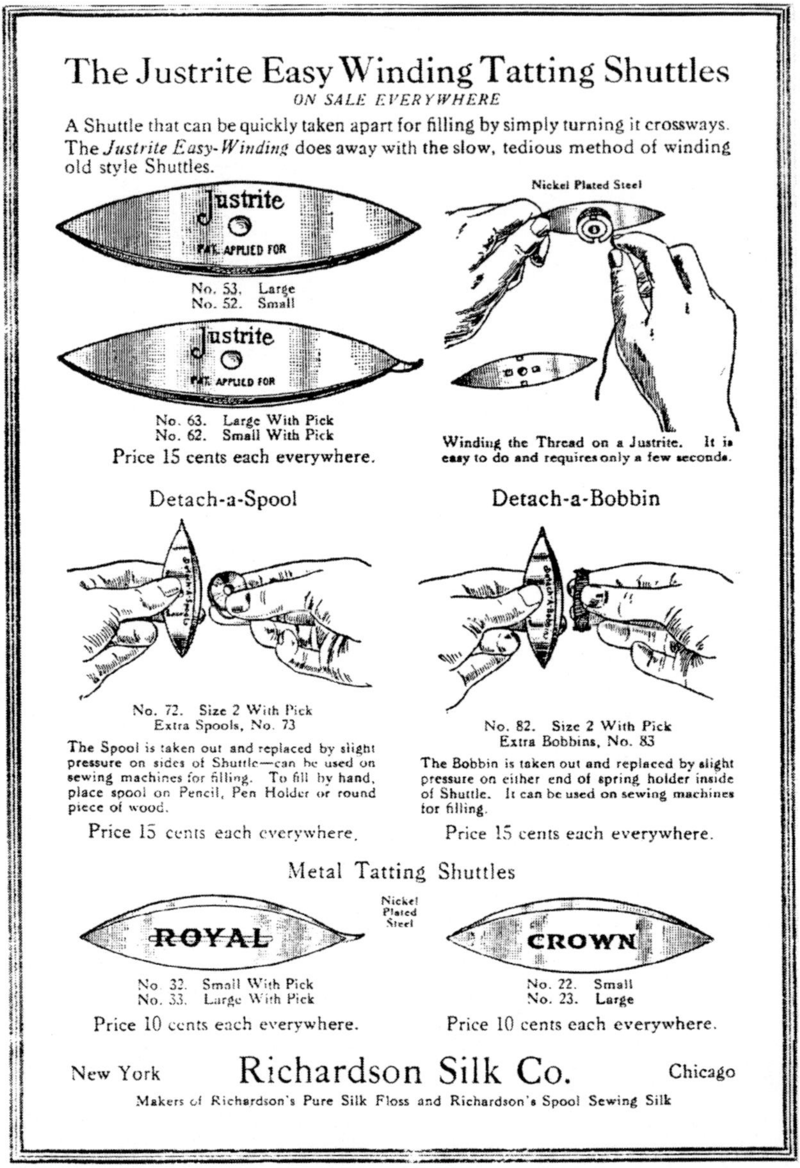

4-1 1916 Richardson Silk Co. advertisment, Tollinger collection.

Chapter Four

Metal Shuttles

Greenwalt

The A. B. Greenwalt shuttle is the first metal tatting shuttle known to be made in America. Patented on January 1, 1867, this lovely shuttle was made of brass. The retractable hook mechanism was originally designed to be on the top, but at final production it was made with the mechanism on the inside. The photographs in figures *4-2* & *4-3* show both sides of this shuttle. One side bears the patent date of "Jan. 1, 1867" bordered by a floral design. The other side has a simple design surrounding the retractable hook. The hook is on a spring that retracts on release of the thumb knob. This shuttle can also be found with a slightly different design on both sides. The Greenwalt patent was also produced later in hard rubber, probably to reduce production costs. At this point

4-2 Front side of Greenwalt shuttle, author's collection.

the hook was changed to a sharp pin. The early hard rubber Greenwalt shuttles have "Pat. Jan. 1, 1867" stamped on one side, the later ones do not. See also the chapters on Rubber Shuttles & Patents.

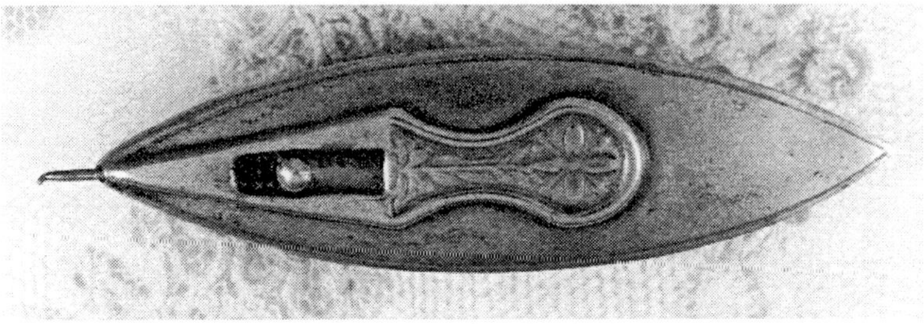

4-3 Retractable hook side of the Greenwalt shuttle, author's collection.

Stockwell

The second American metal shuttle was produced in nickel silver. It was patented by Ira H. Stockwell on August 4, 1868. This patent introduced the improvement of having a stationary point. The small center bump on this shuttle is circled with the words, "Pat. Aug. 4, 1868" bordered by a spray of leaves on each side.

4-4 1868 Stockwell shuttle, author's collection.

Boye

On November 15, 1915 the Boye Needle Company introduced its first tatting shuttles. They had closed ends, one with a small hook, and a removable bobbin on a swing out arm. In the center they were marked "Boye" and "Pat.Appld.For." The shuttles were sold for 25¢ with extra bobbins selling for 10¢ each.

The patent was granted to James H. Boye on April 10, 1917. After the patent was approved, the "Pat.Appld.For" mark was replaced with "Pat. April 10, 1917" on the side of the shuttle. These shuttles were produced until 1922 when they were replaced with the present style patented in 1923.

4-5 Top: 1917 Boye patented shuttle. Bottom: Pre-patented shuttle, author's collection.

The Boye "Improved" shuttle was patented on April 17, 1923 by G. A. Carlson. It was constructed of one piece of folded metal and had a removable bobbin. The folded end is squared off to hold the bobbin for easy winding, the other end had a hook. The shuttle was also designed with three holes on one side, each to give different tension to the thread. This was great in theory, but tatters soon realized that the holes frayed their thread and were not suitable for use. Both this shuttle and its predecessor were made of nickel plated brass until World War II, when zinc replaced nickel for plating. The Boye "Improved" shuttle was sold in two sizes. The larger 3 ½ inch shuttle was first sold in January, 1922 and was discontinued on June 25, 1942. The smaller 3 inch shuttle was sold beginning July, 1922. It was discontinued in January 1943 and then reinstated on February 12, 1947 without the patent date mark. This is the same type of shuttle sold by Boye today.

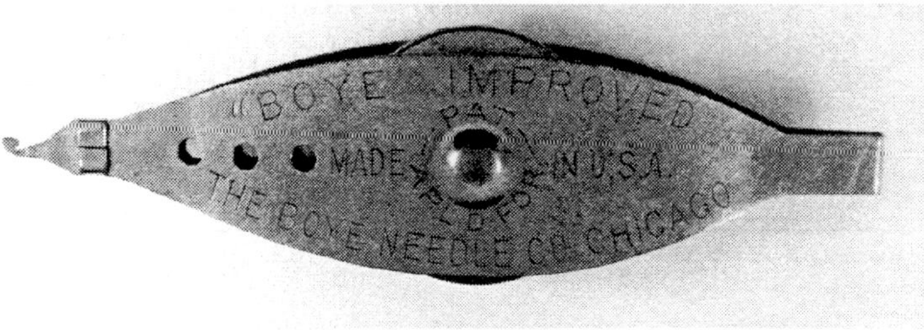

4-6 First 3 1/2 inch Boye "Improved" shuttle, author's collection.

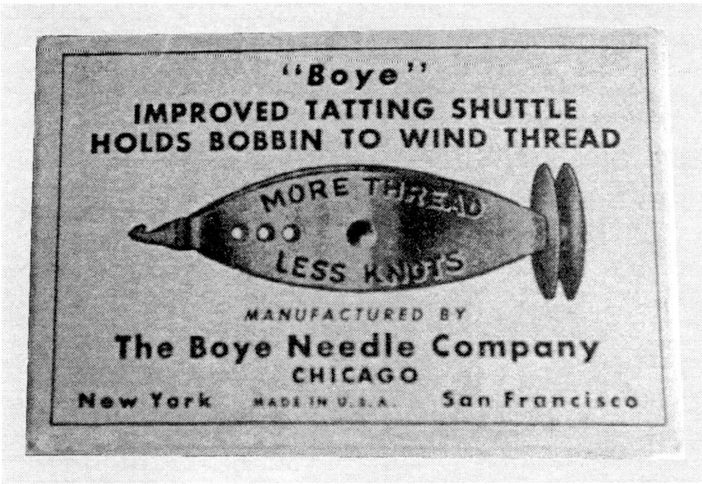

4-7 Original Boye Improved Shuttle box, Woodyard collection.

German Silver

"German silver" is the same as nickel silver and does not contain any true silver. Nickel silver is a metal alloy containing nickel, copper and zinc. From 1915 until the early 1940s most metal shuttles were made of nickel plated brass. During World War II metal shuttle production decreased and producers substituted zinc plating for nickel due to a government order.

One hard to find German silver shuttle is embossed "P.P.I.E. 1915 San Francisco." These shuttle were sold as souvenirs at the 1915 San Francisco Worlds Fair. The Pan Pacific International Exposition ran from February 20 until December 4, 1915. The fair celebrated the completion of the Panama Canal and commemorated the 400th anniversary of the explorer Balboa's discovery of the Pacific Ocean. Shuttles by the same unknown manufacturer may also be found embossed with "California," which was possibly another souvenir of the Worlds Fair.

4-8 German silver 1915 P.P.I.E. shuttle, author's collection.

4-9 German silver marked shuttles, Tollinger collection.

Popular shuttles of German silver were sold by mail order companies around 1917 through the 1920s and were stamped with a decorative filigree design and the words "Ger.Silver." They were sold in two sizes with or without a point. These shuttles are currently being reproduced in the larger pointed size of 3 ¾ inches with either chrome or brass plating by Lacis of Berkeley, California.

4-10 Top: Small German silver without point. Bottom: Large German silver with point, author's collection.

Justrite

In 1916 many metal shuttles appeared on the market. One design was the popular Justrite shuttle patented first by Augie L. Hansen in 1917. This shuttle design had the unique ability to twist apart as the center bobbin was attached to one side making it easy to wind when open. This design was improved with a second patent by Hamitage L. Richardson on May 1, 1923. There are several small differences between the two shuttles, but the most obvious are the 4 square indentations on the back of the 1917 patent in contrast with the two indentations on the 1923 shuttle. There is also an evolution that can be seen in the center hole. The first patent had one that looked countersunk, the next had a straight hole and the last style had a lip around the hole. These shuttles were produced by the Justrite Manufacturing Company of Chicago,

Illinois, and were sold into the 1940s. Today they are being reproduced by Lacis of Berkeley, California and bear the words, "Justrite by Lacis" on one side.

The Justrite Manufacturing Company produced many different metal shuttles, including the Detach-A-Bobbin, Detach-A-Spool, the Royal shuttle and many other plain nickel plated shuttles.

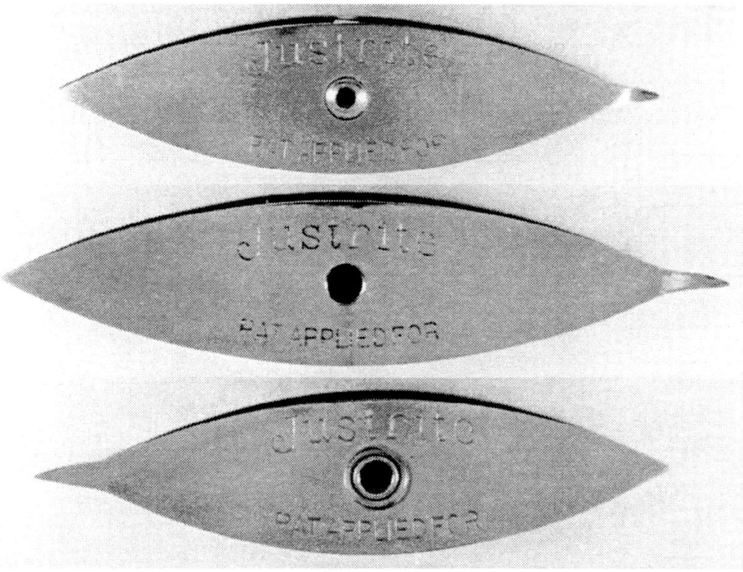

4-11 Top: First 1917 A. L. Hansen patented Justrite. Middle: 1923 Hamitage Richardson patented Justrite. Bottom: Last version of the Justrite.

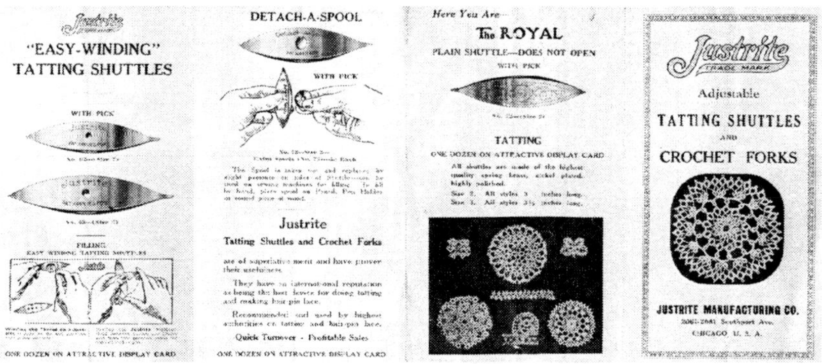

4-12 Undated wholesaler's brochure, author's collection.

Detach-A-Bobbin & Detach-A-Spool

Augie L. Hansen patented a second shuttle design on August 20, 1918. Production of the Detach-A-Bobbin began after the application was filed in December of 1915. This unique design incorporated a sewing machine bobbin. The bobbin could be removed and wound with thread quickly by machine. However, sewing machines bobbins were changed at this time from the long thin bobbin to the short spool type used today. The Detach-A-Spool shuttle was created for the spool type bobbin although no patent was ever filed. Both of these shuttles were produced by the Justrite Manufacturing Company of Chicago, Illinois.

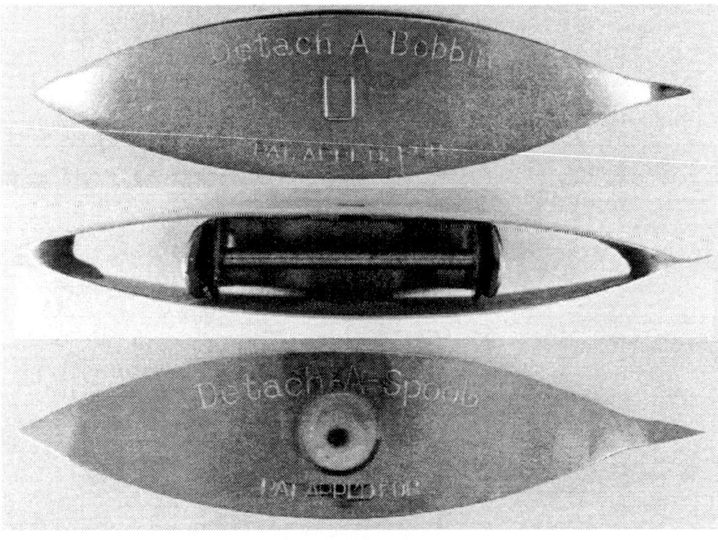

4-13 Top & Middle: Detach-A-Bobbin. Bottom: Detach-A-Spool, author's collection.

Bucilla

In the 1916 *Blue Book of Novelty & Filet Crochet* by Bernhart Uhlman Co. Inc., Bucilla advertised a 2 7/8 inch nickel plated shuttle. These shuttles are actually 2 ¾ inches long. They have "Bucilla" embossed in the middle of both sides, slightly rounded edges, and a small oval post with a center hole. At some later point in time this shuttle was manufactured without the embossed name.

4-14 Early embossed Bucilla shuttle, Demmer collection.

The Royal

Royal shuttles were stamped with an etched-like scroll design on both sides. They were manufactured by the Justrite Manufacturing Company of Chicago, Illinois in two sizes. Advertisements in *Richardson's Tatting Book* of 1916, show the "Royal," a shuttle with a point, and "Crown," a shuttle without a point. Around 1930 the manufacturer changed the center post from a soldered bar to a cylindrical rivet that went though the sides. These shuttles continued to be sold under the name of "Royal" through Bucilla Needlework books by Berhard Uhlman Co. Inc. into the 1940s.

4-15 Top: Original Royal shuttle, large size. Bottom: Later riveted Royal, small size, author's collection.

Unmarked Metal Shuttles

There are two main types of unmarked nickel plated shuttles. They are almost identical, but may be distinguished by two different types of center posts. Both types came in two sizes, 3 inch and 3 1/2 inch, with or without points. One was obviously produced by Justrite Manufacturing Company having the exact same soldered-type center post and sizes as the earliest "Royal" shuttle. The other shuttle has the same center post as the German silver shuttles, comes in the same sizes, and was undoubtedly produced by the same unknown manufacturer.

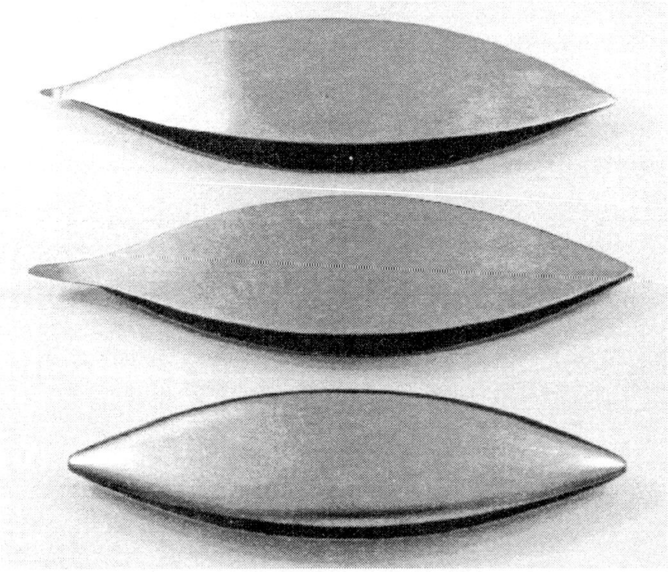

4-16 Top: unmarked shuttle by Justrite. Middle: unmarked German silver type shuttle. Bottom: unmarked Bucilla shuttle, author's collection.

C. D. Fischer

In 1924 Charles D. Fisher was granted an U. S. Patent for his unusual design of a shuttle and winder. On these nickel plated shuttles are the words: C. D. Fisher Mfg. Co., Keyhole Shuttle, and Denver, Colorado. The keyhole name refers to the keyhole shaped slot in the middle of the shuttles in which the accompanying winders fit. Thess shuttle were sold with or without a point until 1932. See also the chapter on Tatting Tools & Kits.

4-17 C. D. Fisher Key Hole shuttle, Demmer collection.

Susan Bates and Marcia Lynn

The "Susan Bates" metal shuttles were produced after 1940 when the Boye patent expired and were made by the C. J. Bates and Sons company. In the 1960s Bates manufactured identical shuttles with no markings marketed as the "Marcia Lynn" shuttles. The company continues to make this same style shuttles today under the Susan Bates name.

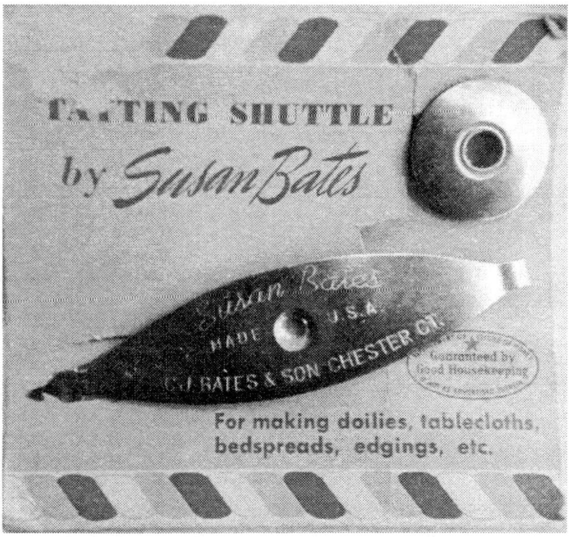

4-18 Susan Bates shuttle on original card, Woodyard collection.

Hero

The Hero Manufacturing Co. of Middleboro, Massachusetts bought the right to produce the Detach-A-Spool and Tilco plastic tatting shuttles as a part of their line of needlework tools around 1948. These shuttles were also stamped with the Hero name. The Hero Manufacturing company was in business until 1973.

4-19 Hero Detach-A-Spool shuttle, author's collection.

Knotting shuttles

CP 1-1 Tortoisesehell inlaid with gold, Middleton collection.

CP 1-2 Mother of pearl inlaid with silver, Woodyard collection.

CP1-3 English silver filigree, Clotilda collection.

CP 2-1 Thomae Co. enamel, Woodyard collection.

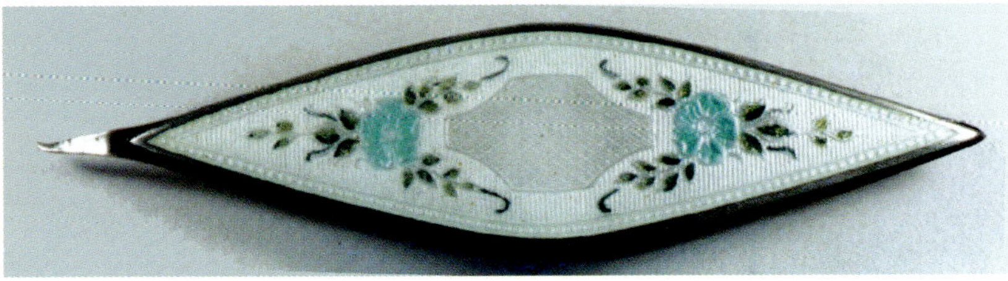

CP 2-2 Webster Co. enamel, also came in pink and lavendar flowers, Woodyard collection.

CP 2-3 Webster Co., Woodyard collection.

CP 2-4 Top: R. Blackington. Bottom: Webster Co., Tollinger collection.

Color plate 2

CP 3-1 N & H, Middleton collection.

CP 3-2 C.B. & H., Middleton collection.

CP 3-3 Variety of solid colored Webster Co. enamels, Tollinger collection.

CP 3-4 R. Blackington enameled sterling shuttle.

Color plate 3

CP 4-1 Webster Co., Middleton collection.

CP 4-2 Unmarked sterling & gold, Middleton collection.

CP 4-3 Leach & Miller, Woodyard collection.

CP 4-4 ALLCO, author's collection.

CP 4-5 Unmarked gold, Demmer collection.

Color plate 4

CP 5-1 Abalone shuttle kits, Top two: Tollinger collection, Bottom: Middleton collection.

Color plate 5

CP 6-1 Abalone shuttles, Tollinger collection.

CP 6-2 Abalone shuttles, Tollinger collection.

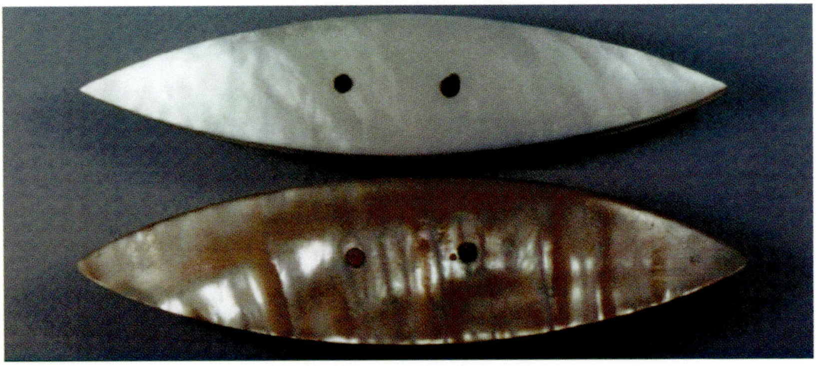

CP 6-3 Comparison; Top: Mother of pearl, Bottom: Abalone, Tollinger collection.

Color plate 6

CP 7-1 Riveted celluliod shuttles, author's collection.

CP 7-2 Beveled edge celluloid shuttles, Tollinger collection.

Color plate 7

CP 8-1 Celluloid shuttle with rounded sides, Tollinger collection.

CP 8-2 Pearlescent celluloid shuttles, Tollinger collection.

Color plate 8

CP 9-1 Two layer celluloid shuttles, Tollinger collection.

CP 9-2 Slit-backed shuttles, Tollinger collection.

CP 10-1 Colorful shuttles of the 1930's, Tollinger and author's collections.

CP 10-2 Susan Bates White Dot shuttles, Tollinger collection.

CP 10-3 Handpainted celluloid shuttle from Pohlson Gifts, Tollinger collection.

Color plate 10

CP 11-1 Tilco & Cook Co. shuttles, Tollinger collection.

CP 11-2 Boye Co. plastic shuttles, Middleton collection.

Color plate 11

CP 12-1 E. B Kingman Co. shuttles, Tollinger collection.

CP 12-2 Shuttles by Merribee Needlecrafts, Tollinger collection.

Chapter Five

Rubber Shuttles

Throughout the world many people have experimented with rubber, the milky white fluid in rubber trees. In America it was Charles Goodyear who had the most success. Early rubber had many objectionable limitations; it became soft in heat, rigid in cold, adhesive with pressure, and had a very unpleasant odor. These characteristics stood in the way of the widespread use of rubber in the early nineteenth century. In 1839 Charles Goodyear discovered that by using the right amount of sulfur along with heat he could produce a vulcanized rubber without the unwanted aspects. He received his first patent for this process in 1843. The product we know today as hard rubber, used to make most shuttles, was granted a U.S. patent in 1851 to Charles' brother, Nelson Goodyear. Hard rubber has also been called vulcanite, and in Great Britain it was known as ebonite because of its resemblance to ebony wood.

In the 1850s there were five companies in America making hard rubber using the Goodyear patent. One of these was the Novelty Rubber Company whose initials, N.R.Co., can be found stamped between the rivets of the first rubber shuttles produced in America. The Novelty Rubber Company was started in 1851 in New York City. It moved to New Brunswick in 1855 and was dissolved in 1884.

The Novelty Rubber Company also produced one of the first U.S. patented tatting shuttles for Constantine Hingher in 1871. Mr. Hingher had been with the Novelty Rubber Company since the company started and had patented many improvements in their manufacturing process. The original wood shuttle Mr. Hingher submitted for patent can been seen by appointment at the National Museum of American History Textiles collection. One side of the hard rubber shuttle produced by the Novelty Rubber Co. bears the words "Hingher Pat. No. 114,008, Goodyear Pat. No. 8075, and N.R.Co. See also the chapter on Patents.

5-1 Original Hingher patented shuttle, courtesy of the National Museum of American History, Textile collection.

Adolf Dobrowsky of Shasta, California chose hard rubber for the production of his 1866 patent. The patent called for a "pin passing through the shuttle longitudiually" with both ends having a point. However, in the final production of these shuttles the pin only goes through half of the shuttle. These shuttles were stamped with the words "Pat. Appld. For."

5-2 Dobrowsky patent, Middleton collection.

An unusual shuttle made of rubber with a retractable pin mechanism was patented by A. B. Greenwalt in 1868. Originally made of brass this shuttle was presumably changed to hard rubber to reduce production cost. Early hard rubber shuttles bear the "Pat. Jan. 1, 1867" patent date, however, later ones do not.

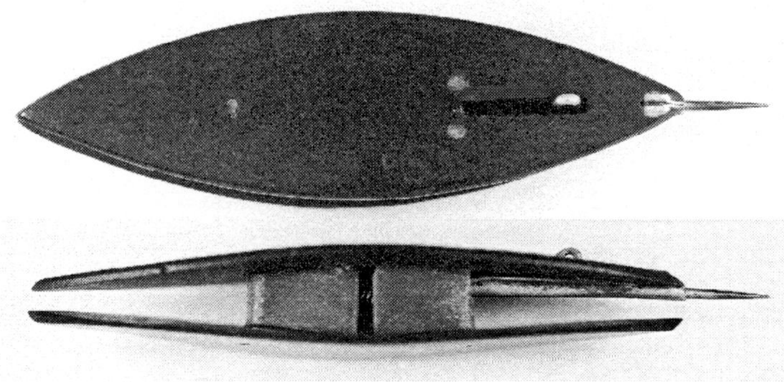

5-3 Greenwalt patent in hard rubber, Tollinger collection.

Along with the Novelty Rubber Company, the Goodyear Company was also manufacturing shuttles in the late 1800s. By the twentieth century from 1914 to 1940 the American Hard Rubber Company was the primary producer of rubber shuttles.

TATTING SHUTTLES.

No.	Per gro.	No.	per gro.
1	$10 00	3	$15 00
2	12 50	4. Patent	30 00

5-4 Goodyear Rubber Co. 1878 Trade Journal, courtesy of U.C. Santa Barbara.

Hard Rubber Tatting Shuttle. Price 20 cts. postpaid.

5-5 1898 Ladies Art Co. publication, courtesy of the Missouri Historical Society.

Gutta percha is a close relative of rubber. However, the gum comes from a different species of tree and is different chemically because it contains oxygen. This difference is significant because oxygen made it impossible for gutta percha to harden and resistant to the action of sulfur and other corrosive substances. This quality made gutta percha important as a telegraph cable insulation. If gutta percha was used for shuttles, it would have been mixed with rubber so it could be hardened. This author has not found evidence of any gutta percha company manufacturing tatting shuttles.

To identify a shuttle as being made of rubber, rub it briskly with a finger, the warmth of the fiction will create the same odor as a burnt tire. It is important not to use any liquids on rubber shuttles. Water will turn a black rubber shuttle to an olive green color. Some rubber shuttles have turned a mottled yellowish color due to the sulfur content. Sulfur was used as a drying agent for rubber and has caused many rubber shuttles to become brittle and brown. Rubber shuttles may be cleaned with simple hand rubbing which allows natural skin oils to bring out a shine.

The Julia Shuttle

"The Julia Shuttle" was the name given to an interesting modification of rubber shuttles. The suggestion was illustrated in the 1898 magazine *The Designer*. One could take an ordinary black rubber shuttle, cut off one end slightly and whittle the opposite side to a point. This was said to improve the ability to join tatting, thus saving time by eliminating the use of a pin.

THE DESIGNER. SEPTEMBER, 1898.

Knitting and Tatting.

THE TATTED JULIA MEDALLION.—Make 6 d s, 1 p, 3 d s, 1 p, 3 d s, 1 p, 6 d s· draw up and repeat 8 times and tie. Join last p to first p after tying. 3 p, draw up, 3 p, draw up. For outside loops 6 d s one-half inch p before draw up, 2 d s until there are 7 purls, 6 d s, draw up, 3 d s, draw up, 4 d s, join in first p of the wheel of 8 loops, 4 d s, draw up. Repeat as in illustration until there are 16 loops on the outside joined together. Have the purls long and use medallion to set in corners of centre-piece, etc., by being button-holed around edge on cloth, and then cut cloth from under.

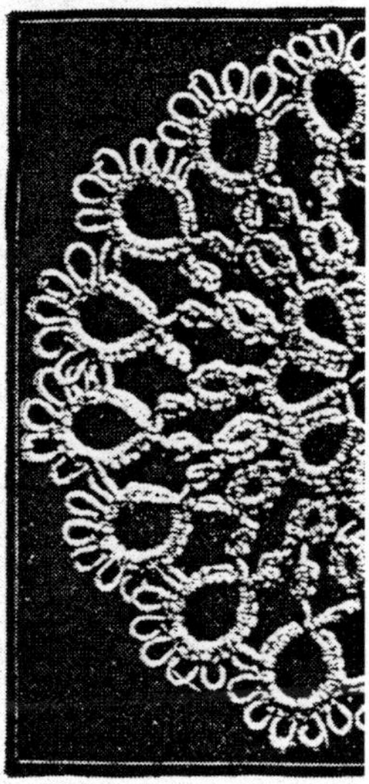

THE JULIA

The Julia Shuttle.—Take an ordinary black rubber shuttle and with a sharp penknife cut one point off as shown in No. 2. The opposite point of the same end of shuttle should be whittled to a point as in No. 1. By so shaping the shuttle a pin will not be found necessary when joining the tatting. Anyone who has once used a shuttle of this description will find it a great improvement, and will save much time by its employment.

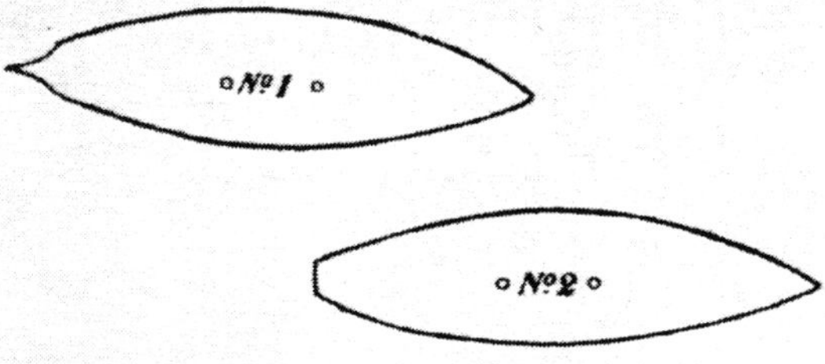

THE JULIA SHUTTLE.

5-6 1898 *Designer* magazine, courtesy of the National Museum of American History, Archives collection.

6-1 Celluloid shuttle advertisement in a 1917 Butler Bros. catalogue, author's collection.

Chapter Six

Plastic Shuttles

More shuttles were made of plastics than of any other material. Plastic by definition is any material that can be molded or shaped. This broad definition includes a wide range of materials from natural substances like horn and ceramics to man-made synthetics. This chapter focuses on the various man-made plastics used to make tatting shuttles.

In the middle of the nineteenth century scientists and inventors all over the world were experimenting with various materials to find a substitute for ivory, including the very flammable cellulose nitrate, the main ingredient of gun powder. It was the well publicized 1863 challenge of Phelan & Collender, offering a $10,000 reward to anyone inventing a substitute for the ivory used in billiard balls, that inspired the American John Wesley Hyatt to try his hand. Hyatt was not only successful in developing and refining the right combination of cellulose nitrate and camphor to make Celluloid™, but also for making it a commercially successful product. The Celluloid Manufacturing Co. was formed in 1871, however, this author has found no evidence of celluloid shuttles being made before 1905.

The first celluloid shuttle appeared in a 1905 advertisement of sewing notions in the Spring and Summer catalogue of John Wanamaker of Philadelphia. The riveted shuttle design was offered in black and white and, as with many shuttles, the manufacturer is unknown. The majority of celluloid tatting shuttle advertisements are found from 1910 to 1940 when tatting was at its highest popularity. Around 1915 colors other than black and white were introduced; including pink, blue, faux tortoiseshell and French ivory. French ivory was the name given to the celluloid made with light and dark colored parallel lines to resemble natural ivory. Celluloid and other cellulose plastics are known as thermoplastics. Thermoplastic material becomes soft and moldable under heat or pressure. From the turn of the century until 1945 celluloid was the most common plastic used in America.

Celluloid's high flammability property stimulated development of non-burning plastics such as celluloid acetate, phenolic plastics and many others. Celluloid acetate is also a thermoplastic but was not often used in making shuttles. Bakelite, the first thermosetting plastic, was developed by Leo Bakeland in 1909. Thermosetting plastic is characterized by its ability to be molded when heated but since the process is accompanied by a chemical change, it is not reversible. Thus bakelite shuttles never lose their shape nor do they become sprung. It wasn't until the 1930s, well after the patent expired, that this material became widely used, although it was not often used for shuttles.

It is possible to test a plastic shuttle with a hot needle on the inside to determine whether it is celluloid or bakelite, but one should be aware that this does leave a mark on most shuttles. Celluloid smokes and melts immediately when heated and will have a strong smell of camphor. Camphor is the ingredient in Vic's Vaborub that gives it a very medicinal odor. Bakelite on the other hand does not burn, smoke or melt. For those who would like more information on this subject, many of the button collecting books are excellent sources.

The majority of plastic shuttles were unmarked and the dates of manufacture can only be approximated. However, there were several marked shuttles produced as advertisement gimmicks or used in marketing products, services, and retailers.

Collingbourne

The first dateable plastic shuttles were sold by Collingbourne in their "Encyclopedia of Technologic Art Needle Work Instruction" book by Virginia Snow dated 1915. These unusual flat shuttles bearing the name "Collingbourne" came in white, black and tortoise shell color and were produced in two sizes. The smaller shuttle was produced first with the words, "Pat. Pend." before the patent date. They were sold as "the rapid shuttle" because they were easy to wind.

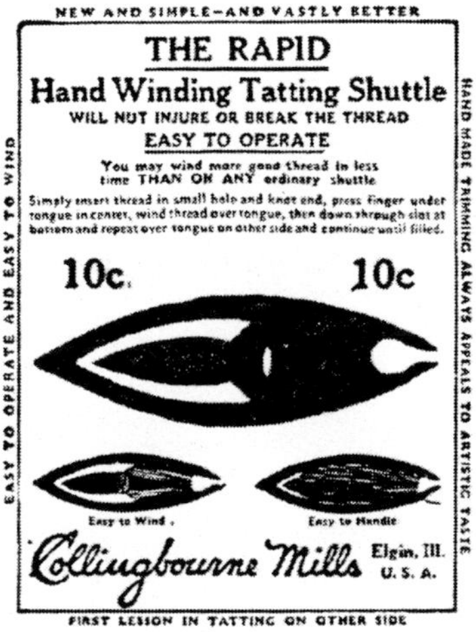

6-2 Large shuttle 3 1/2 inches, Tollinger collection, small shuttle 2 5/8 inches, Hollingsworth collection.

Bucilla

Other marked plastic shuttles that were offered with a needlework manufacturer's name were stamped with "Bucilla" followed by the words "Reg.U.S.Pat.Off." According to the 1916 *Bucilla Blue Book of Novelty & Filet Crochet* these celluloid shuttles were made in white, black and tortoiseshell colors. They came in sizes 1, 2, 3, 4, 5, (5 is the largest at 4 inches long). Pink and blue shuttles were available in size 2 only. A 1940 advertisement shows Bucilla offering three unmarked celluloid shuttles, a 3 ½ inch with a point in white or black, three sizes without points ranging from 2 ¾ inches to 4 inches, and 3 ½ inch with a hooked end in black or white.

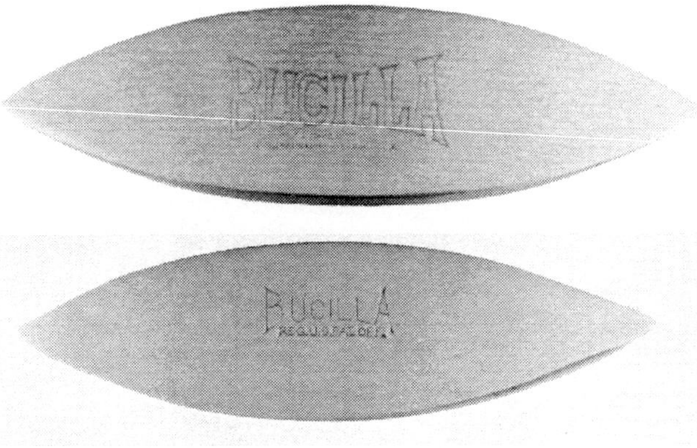

6-3 Bucilla shuttles, Tollinger collection

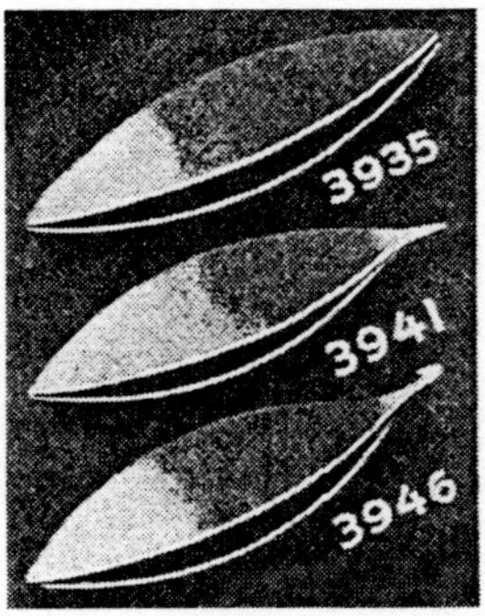

Art. 3935—Ivory Celluloid. White or Black, 1 dozen in a box.
Size 2—2¾x¾" .... **$ .85 Doz.**
Size 3—3¼x¾" .... **.90 Doz.**
Size 4—4x1" .... **1.10 Doz.**

Art. 3941—With Pick End Ivory Celluloid, size 3½x¾". White or Black only. 1 dozen in box.... **$.75 Doz.**

Art. 3946—With Hook End Ivory Celluloid, size 3½x¾". White or Black. 1 dozen in box........ **$1.00 Doz.**

6-4 Bucilla Fall & Winter 1940 catalogue, courtesy of the National Museum of American History.

Goddard

The shuttle marked "Goddard Shuttle" is slightly different than C.F. Dill's 1916 U.S. Patent, but fits most of its description. The sides have small round depressions to hold the bobbin in place. These shuttles are marked "Pat. Pend." and were probably produced as early as 1914 when the patent application was filed. The Goddard was produced in two sizes mostly in French ivory, but occasionally they are found in colors. Around 1917 a similar shuttle was made with the ends riveted together and the words "Pat. Pend." stamped on the shuttle.

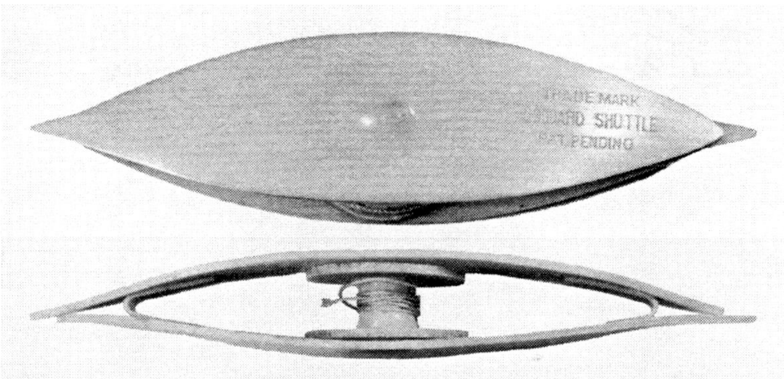

6-5 Goddard shuttle front & side views, Tollinger collection.

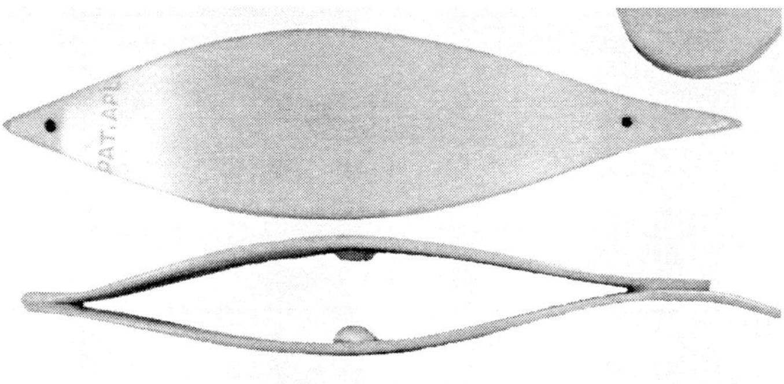

6-6 Shuttle with riveted ends & removable bobbin, Tollinger collection.

Lydia Pinkham

Currently this is the most sought after advertising shuttle and it was once given away free to respondents who would fill out a questionnaire. The questionnaire appeared in the Lydia Pinkham publications of 1917 and 1921. Lydia Pinkham was well known for her cure-all tonic, which was marketed to women and had a high alcohol content. These French ivory shuttles made of celluloid, pictured an image of Lydia Pinkham on one side and her name on the other, along with the words "Yours for Health" and "Vegetable Compound." In small print along the pointed end of the shuttle is the manufacturer's name, "The Whitehead & Hoag Co., Newark, N.J." This large company was in business from 1890 to 1935 and quite possibly manufactured tatting shuttles for other companies.

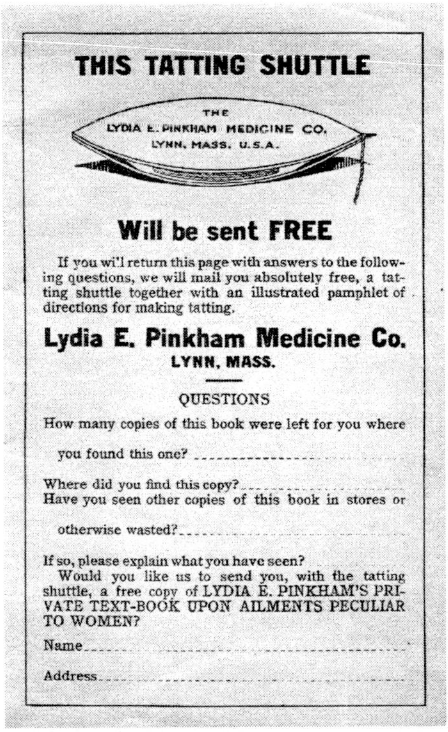

THIS TATTING SHUTTLE

THE LYDIA E. PINKHAM MEDICINE CO. LYNN, MASS. U.S.A.

Will be sent FREE

If you will return this page with answers to the following questions, we will mail you absolutely free, a tatting shuttle together with an illustrated pamphlet of directions for making tatting.

Lydia E. Pinkham Medicine Co.
LYNN, MASS.

QUESTIONS

How many copies of this book were left for you where you found this one? ....

Where did you find this copy? ....

Have you seen other copies of this book in stores or otherwise wasted? ....

If so, please explain what you have seen? ....

Would you like us to send you, with the tatting shuttle, a free copy of LYDIA E. PINKHAM'S PRIVATE TEXT-BOOK UPON AILMENTS PECULIAR TO WOMEN?

Name ....

Address ....

6-7 1917 Lydia Pinkham publication "Facts and Fancies," author's collection. Right: Lydia Pinkham celluloid shuttle, Tollinger collection.

6-8 Lydia Pinkham celluloid shuttle, Tollinger collection.

Inlaid Co.

David Butler, assignor of the Inlaid Company, Inc. of Providence, Rhode Island received a patent for his improved tatting shuttle in 1918. The sides of these shuttles have indentations that form its center post. Embossed in this center indentation are the words "Pat. Pend." These shuttles were probably being sold in 1917. The outside edge of the shuttle has a flat rim to impart rigidity to this very thin celluloid shuttle. In a 1920 Butler Bros. Catalogue, a shuttle is pictured without the middle indentation but with similar flat rimed sides. These shuttle were sold in four different sizes and came in popular colors of the era: black, ivory, blue and pink. Today they are affectionately called the pumpkin seed shuttles.

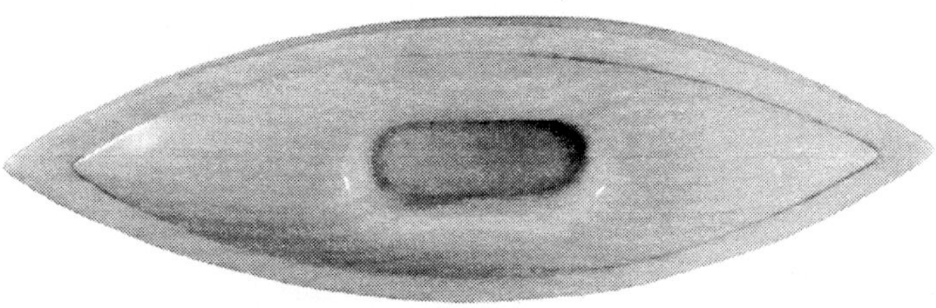

6-9 French ivory Inlaid Co. Shuttle, Middleton collection.

6-10 Pumpkin seed shuttle of the 1920s, author's collection.

There are two other shuttles that could have been either copies of or evolutions of the Inlaid patent. One was made of a heavier celluloid with indented sides that form a center post and was riveted in the center. The other is one has two ¾ inch slits in either side that indent to form an inner post.

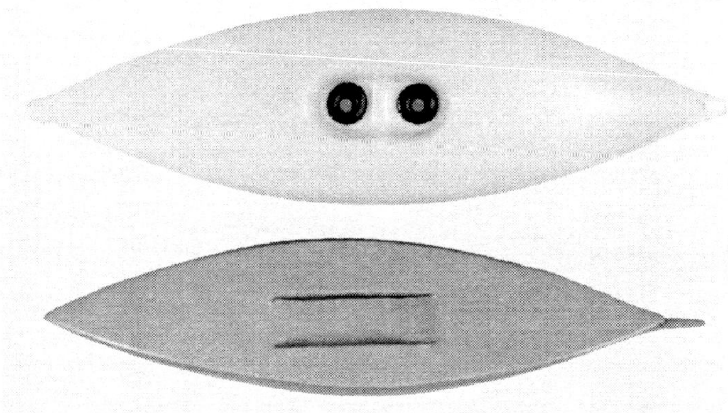

6-11 Shuttles with indented sides, Tollinger collection.

The shuttle made by the Inlaid Co. was also sold in a popular catalogue of the 1920s called "Pohlson Gifts." It was advertised as an ivory shuttle (there was no truth-in-advertising law at this point in time) along with a carrying case and a ball of thread. Although Pohlson Gifts of Pawtucket, Rhode Island didn't manufacture shuttles, tatting shuttles as well as other items they sold may often be identified because of the addition of hand painting on their items. Besides popular floral designs, little blue sparrows were often painted on many objects they sold. The Pohlson Gifts company was also known for the sweet little poems that were printed on the accompanying gift boxes or cards. See color plate 10, figure *CP 10-3*.

6-12 Pohlson Gifts Advertisement, courtesy of National Museum of American History.

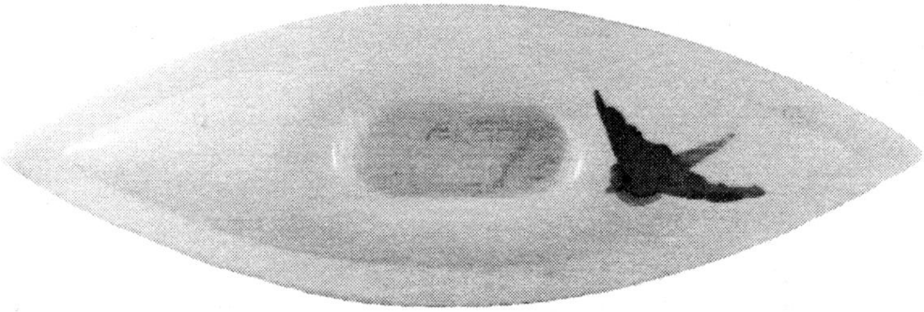

6-13 Inlaid Co. celluloid shuttle with typical Pohlson Gifts hand painted swallow, author's collection.

Carnation

Other popular giveaway shuttles of the 1920s were ones marked "Carnation" and "trademark." These shuttle were given out as an advertising gimmicks in the 1920s to promote the Carnation Evaporated Milk Co. The shuttles came in two sizes without points, in white, black, tortoiseshell, pink and blue.

6-14 Shuttle marked Carnation Trademark, Tollinger collection.

ABCo

The Asmus Bradley Co. of Chicago was another needlework supply retailer that put its name on the shuttles they sold. The shuttles are marked with the initials ABCo encased in a diamond centered on the middle of one side. The Asmus Bradley Co. can be found listed in the business directories from 1924 to 1945.

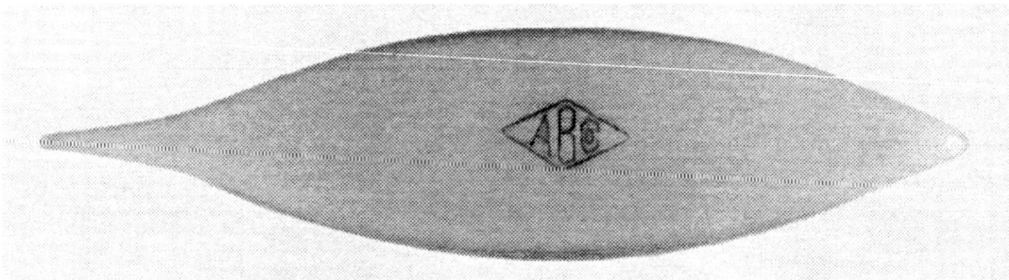

6-15 Asmus Bradley Co. shuttle, Tollinger collection.

Susan Bates White Dot

C. J. Bates, best known for making bone shuttles before the turn of the century, began selling celluloid shuttles around 1917. These unmarked shuttles, usually in French ivory, were found in the later leather sewing kits in which C. J. Bates originally sold their bone shuttles. The C. J. Bates company sold a wide variety of sewing kits with crochet hooks, knitting needles and other sewing accessories as well as manicure sets. It is unknown at this time if C. J. Bates actually produced its own celluloid products. In 1943 and 1944 Louis A. Stillman, as assignor of the C. J. Bates Company, patented two tatting shuttles. Both of these patents were being produced at the same time, the basic difference is that one had a point and the other did not. These shuttles had a raised rim around a white center depression on either side where the center post is connected. This was bordered by the name "Susan Bates" on top, a brand name adopted by C. J. Bates in 1937, the words "White Dot" on either side of the central depression and "Pat. App'd For" on the bottom. These shuttles were made in white, black, blue, green and red. They were also produced with white marbling in red, blue and green.

6-16 Susan Bates White Dot shuttle, author's collection.

Colorful shuttles of the 1930s

In the 1930s there was an explosion of color in shuttle production. Not only were colors ranging from clear amber to any color imaginable, but the most outstanding also had inlaid designs with stripes, confetti, marbling and snakeskin patterns. These shuttles were sold on paper display cards bearing the words "Tatting Shuttle," "Finest Quality," "Made in U.S.A.", and "Keep away from artificial heat." Although any heat could have harmed these celluloid shuttles. In the years from 1933 thru 1935, these cards were printed with an eagle symbol and the initials N.R.A. This represented the National Industrial Recovery Act created by the National Recovery Administration to enforce fair business practices during the American depression. When compared to the Susan Bates "White Dot" shuttle, there is an unquestionable similarity, leading to the assumption that these shuttles were made by the same company. There was another similarity in the marketing. The shuttle display cards were almost identical to those of the Tilco & Cook Company of Leominster, Mass.

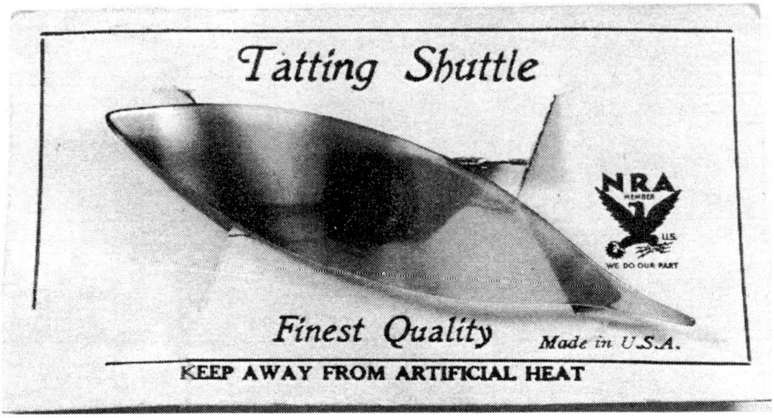

6-17 1930s shuttle on card marked with NRA, Woodyard collection.

Tilco

The Tilco & Cook Co. began about 1880 making combs, jewelry, hairpins, bracelets and novelties from horn. By 1905 they were listed in Massachusetts' directories as manufacturers under both "Horn Goods" and "Celluloid Goods." Around 1945 they made and sold a molded thermosetting plastic shuttle on a display card with the words "Tatting Shuttle," "A Tilco Product" and "Made in the U.S.A." In 1951 this same shuttle was sold with the marking "Hero" by the Hero Manufacturing Co. of Middleboro, Massachusetts and continued to be sold into the 1960s. Because Tilco & Cook made shuttles to be sold by other companies and marketed under other names, it is very possible that Tilco & Cook made the "Susan Bates White Dot" shuttles. Whether the earlier colorful shuttles of the 1930s that were also sold on display cards were manufactured by the Tilco & Cook Co. or by C. J. Bates is unknown at this time.

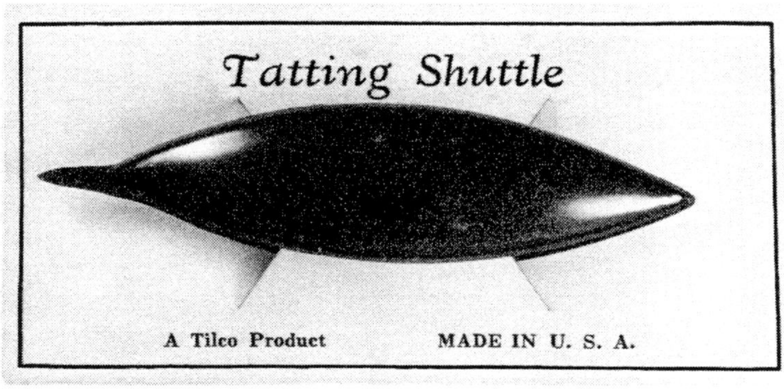

6-18 Tilco shuttle on original card, author's collection.

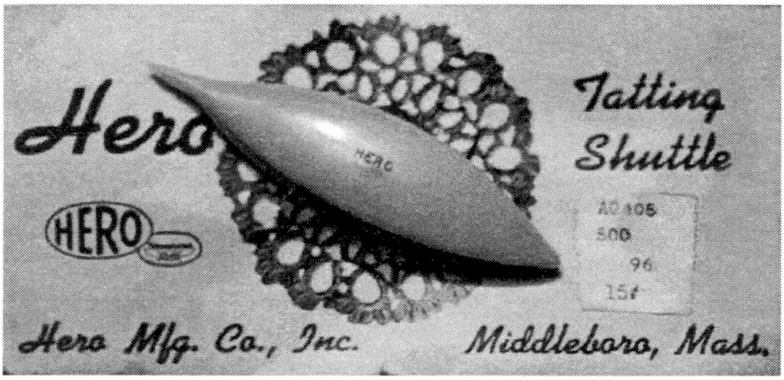

6-19 Plastic Hero shuttle, Stevens collection.

Gloria

The name "Gloria" was held as a trademark of the McCoy & Jones Co. of Chicago from approximately 1928 to 1945. This company was a notions distributor and manufacturer of "Gloria Pins." On shuttles the name "Gloria" appears in cursive in the middle of the word "Trademark," usually on one side of the shuttles. These shuttles are generally found in white, black, tortoiseshell or French ivory in two sizes. The smaller shuttles also came with or without a point. The larger ones were 3 ¼ inches long and unusually wide at 1 inch. The resemblance of the "Gloria" mark and the "Carnation" mark lead one to suspect that these shuttles may have been made by the same unknown manufacturer.

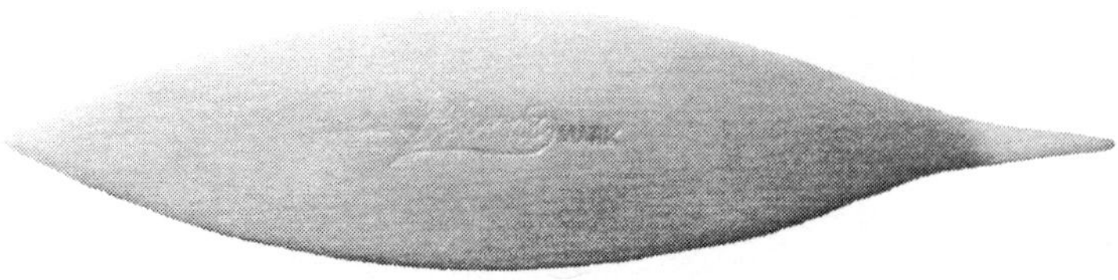

6-20 Shuttle marked Gloria Trademark, author's collection.

Boye

The Boye Company's first plastic shuttle was an uncommon flat shuttle patented by G. A. Carlson. They were introduced in June of 1947 and produced in four colors: red, yellow, green and white. These shuttles were discontinued in May of 1952 when Boye's present-day plastic shuttles were introduced in the same colors.

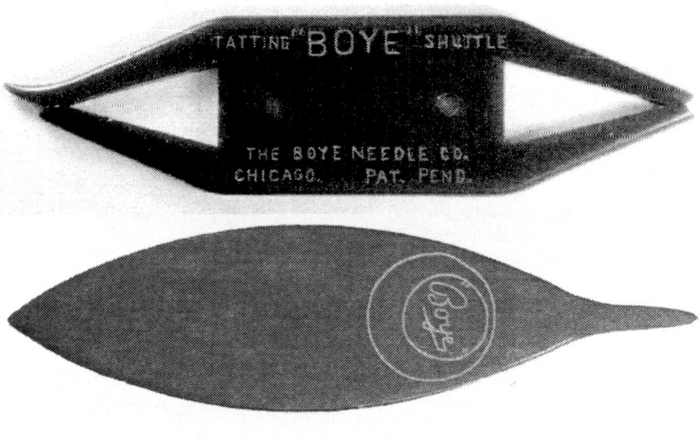

6-21 Top: First plastic Boye shuttle. Bottom: Second and present Boye plastic shuttle, author's collection.

Early unmarked shuttles

Future research will hopefully uncover other notions catalogues that may help to identify makers and dates of some of the more distinctive shuttles of yet unknown origins. For example one such shuttle is an early celluloid pierced with a 1920s art deco design. The unknown maker of this shuttle also made a large plain shuttle sharing the same close rivet location and large oval center post. These shuttles are found in white, black, blue and pink.

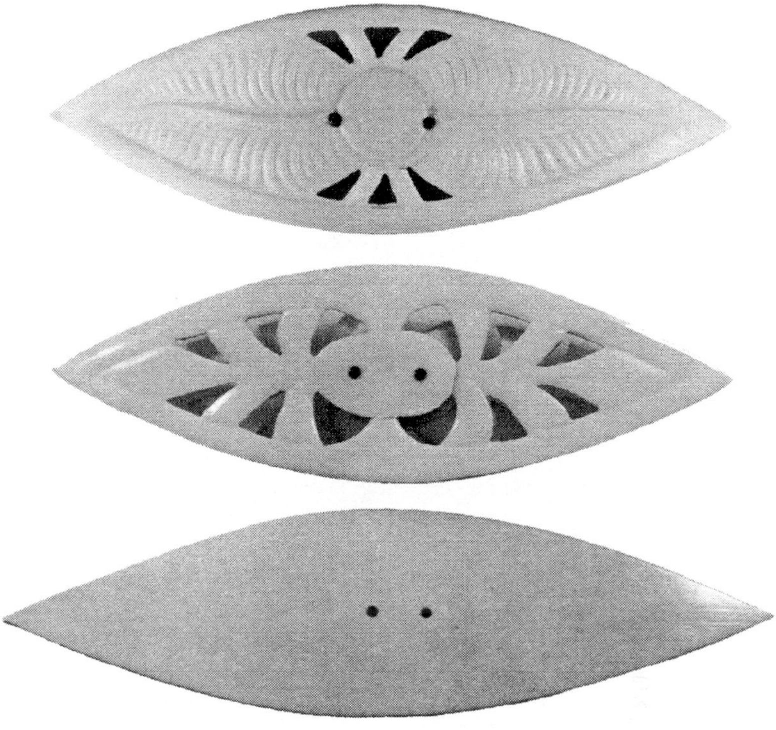

6-22 Pierced celluloid shuttles, Middleton & author's collections

Another popular shuttle style of the 1920s was of simulated pearl. Originally this effect was achieved by adding millions of tiny flakes from the surface of fish scales to the liquid plastic. In some countries ground mica-coral and marble were added to the material to achieve this effect. When this process became too expensive, a chemist discovered the essential ingredient that served as a substitute. This was the tiny, flake-like crystals of mercurous chloride; household mercury, suspended in cellulosic solution. The earliest of these shuttles often came with a second layer of celluloid usually in a contrasting color. Vanity sets produced by the E.I. duPont company were often found made in this style. The Viscoloid Co. was listed as a shuttle maker in the *Thomas Register of Manufacturers* in 1922 and was acquired by the E. I. duPont Company in 1925. It is quite possible that these simulated pearl shuttles were produced by the Viscoloid Co. See color plate 8, figure *CP 8-2*.

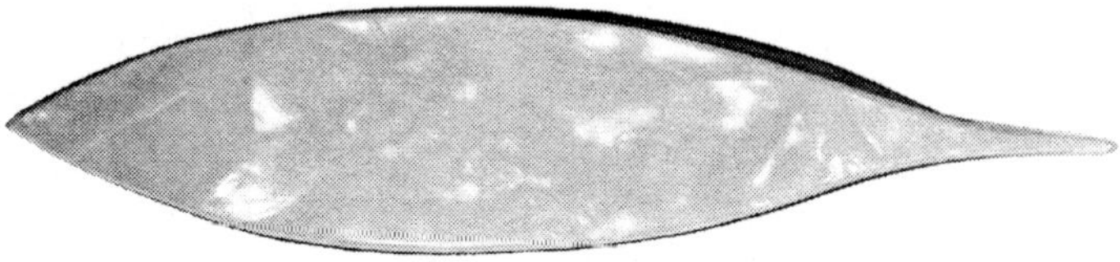

6-23 Pearlescent celluloid shuttle, Middleton collection.

E. B. Kingman Co.

The E. B. Kingman Co. and its subsidiary company, Ben Hur of Leominster, Massachusetts, were celluloid manufacturing companies listed in the *Thomas Register Of Manufacturers* as producing tatting shuttles from 1925 to 1957. However, the E. B. Kingman Co. was in business for over 100 years, from 1888 until 1999. The shuttles clearly linked to this company were made of a thermosetting plastic and marked on the inside with "K - U.S.A." Due to its longevity, the E.B. Kingman Co. could be responsible for many of the unmarked tatting shuttles found today.

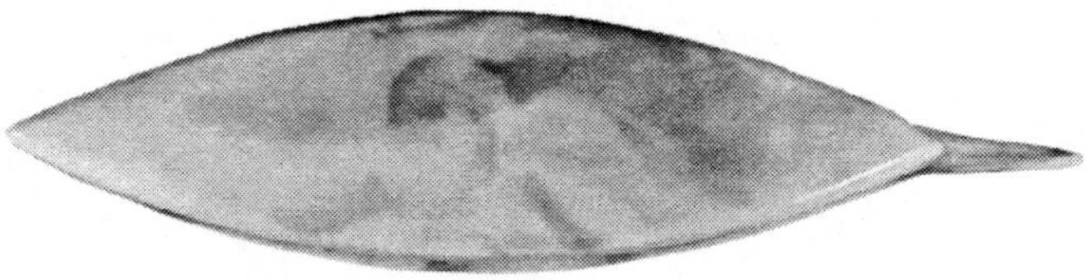

6-24 Plastic shuttle marked "K - U. S. A.," author's collection.

English

English shuttles commonly found in American collections include those made by Milward and Aero. The Milward company made two different types of shuttles. The first was a basic black

Bakelite shuttle which probably sold from the 1930s to 1940s. The second was a black and white plastic shuttle with the center of the sides impressed to form a post. These shuttles were sold on cards with small crochet hooks from the late 1950s until the mid 1980s. Aero's first shuttle was made from aluminum and went out of production around the mid 1970s. Their second shuttle design was made of a gray plastic with a crochet hook attached to one end with a detachable black plastic bobbin. This shuttle is no longer made in England but can be found today. A few other unmarked varieties of English shuttles have made their way to America and many of these were made of thermosetting plastics.

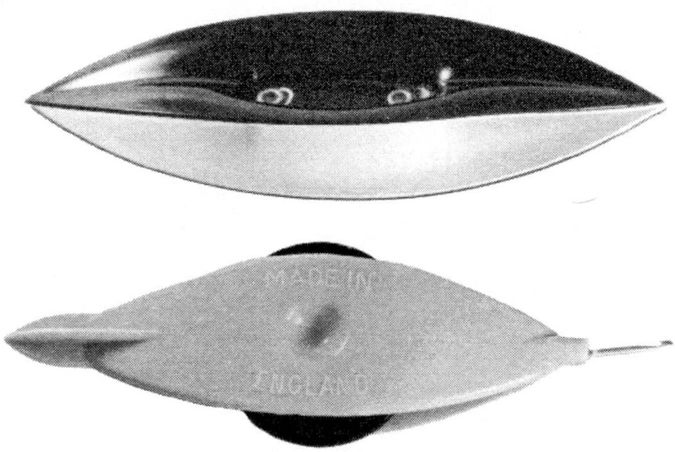

6-25 Top: Milward shuttle. Bottom: Aero shuttle, author's collection.

Contemporary Plastic Shuttles

Clover

The C. J. Bates Company began selling its present day, small plastic shuttle in the early 1970s under the Susan Bates name. This same shuttle was sold by Clover Mfg. beginning in the late 1980s packaged in pairs in a tortoiseshell color. In the late 1990s

Clover began packaging five shuttles for sale together in the neon colors of pink, orange, green, yellow and blue.

6-26 Top: Susan Bates shuttle. Bottom: Clover shuttle, author's collection.

Tatsy

Carol Winandy's first plastic shuttles came out in 1975 - 1976. They were made of a smoky colored clear acrylic and carried a "Tatsy" sticker. The shuttles were hand formed in three sizes: 6 inches, 4 inches, and 2 ½-3 inches. Later shuttles were made in a beryllium alloy mold in different colors embossed with the "Tatsy" name. The present shuttles, made in a steel mold, were originally distributed nationally by Ben Franklin Crafts embossed with the Tatsy logo. In November of 1990 the license was sold to C. & C. of North America and in 1992 the license was returned to Carol Winandy.

6-27 Contemporary Tatsy shuttle, author's collection.

Merribee

In the mid 1970s Merribee Needlecrafts, a subsidiary of the Tandy Corporation, sold small plastic tatting shuttles. They had rounded sides with a small round center post and a short point. The shuttles came in four colors: faux tortoiseshell, white, blue and pink. See color plate 12, figure *CP 12-2*.

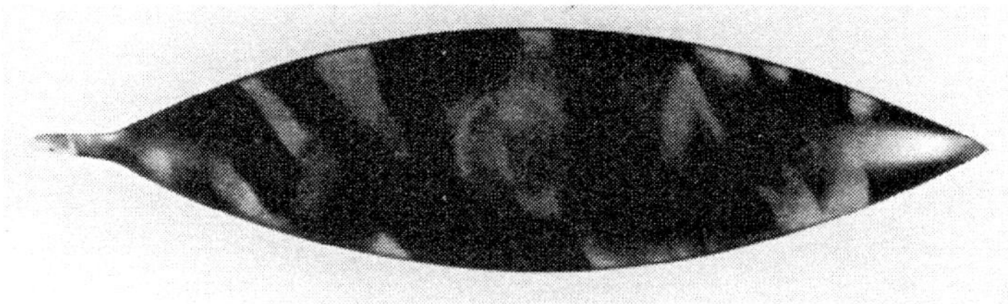

6-28 Merribee shuttle, Tollinger collection.

Lael

Lael Morgan received a U.S. Patent on a shuttle she designed in 1977. The shuttle was ideal for thicker thread and yarn. According to Amos Hicks of Illinois, who manufactured this cast molded plastic shuttle, between 75,000 and 80,000 were produced. These soft plastic shuttles are 5 ½ inches in length and made to resemble wood. The marketer, Hultgren & Lukes, subcontracted with Lois Moyer to sell the shuttles in kits containing designs for tatted snowflakes by Ms. Morgan. By 1980 they were the second largest sellers in Hershner's catalogue and had also sold very well in *Better Homes & Gardens* magazine.

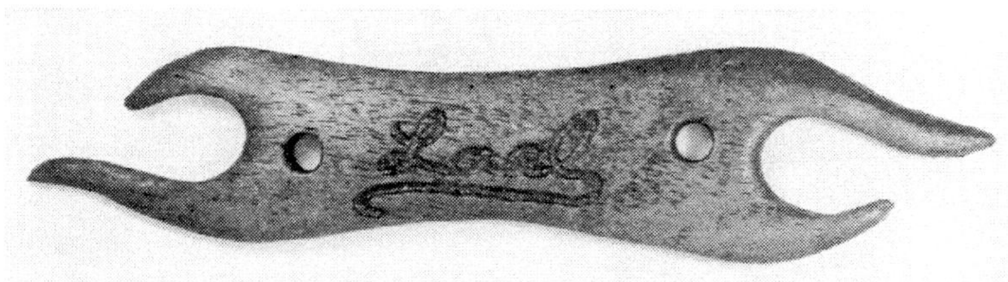

6-29 Lael shuttle, author's collection

LoLo

In 1982 Lois Moyer designed a slightly different flat shuttle which she called the LoLo shuttle. These shuttles were 4 inches long and 1 3/8 inches wide. In the first production, the LoLo shuttles were black painted wood with a LoLo sticker in the middle. In the final production, the shuttles were made in black plastic, also with a LoLo sticker.

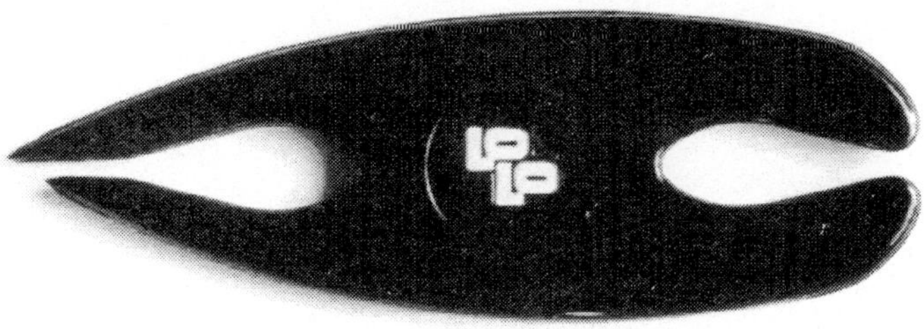

6-230 LoLo shuttle, Alderson collection.

Shuttlebug

From 1981 until 1998 Carole Jackson sold shuttles similar to those of Lael Mogan design. Jackson's shuttles were made of a hard white plastic and trademarked with the name "Shuttlebug" embossed on the sides. They were marketed by Carousel Crafts of Houston, Texas and also sold with thread and tatted snowflake designs.

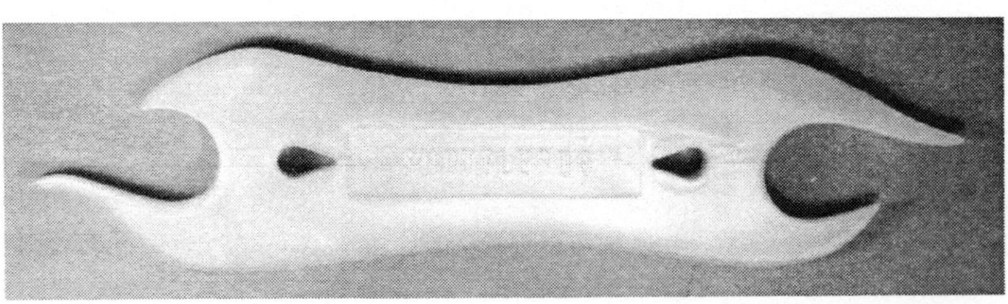

6-31 Shuttlebug shuttle, author's collection.

Debbie-A-Comfi

In 1997 Debbie Arens' tatting shuttle design was patented. The shuttle is molded into one integrated part, has raised edges on the sides and a deep grid for grip. Additionally the Debbie-A-Comfi shuttles were designed to fit a more natural finger curve. Ms. Arens began selling her shuttles in blue and white while the patent was pending in 1996. They are now being produced in three additional colors: teal, pink and lavender.

6-32 Debbie-A-Comfi shuttle, author's collection.

Jewelry

Lauren Snyder is best known for her lace tool designs in jewelry made of sterling silver. She began carving full sized shuttles in 1997 from cast acrylics such as Corian. She has also produced these in miniature for jewelry. Additionally, she has made a few tatting shuttles from countertop laminates.

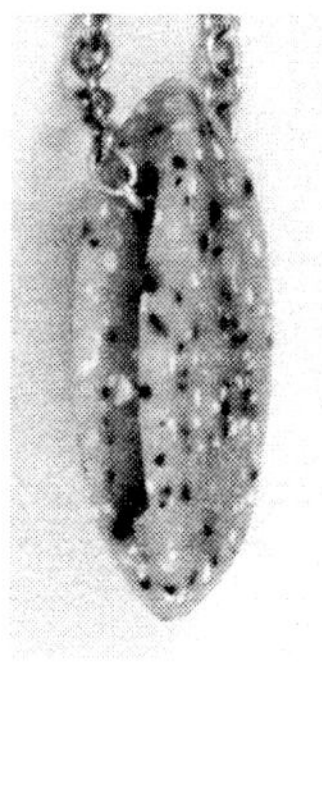

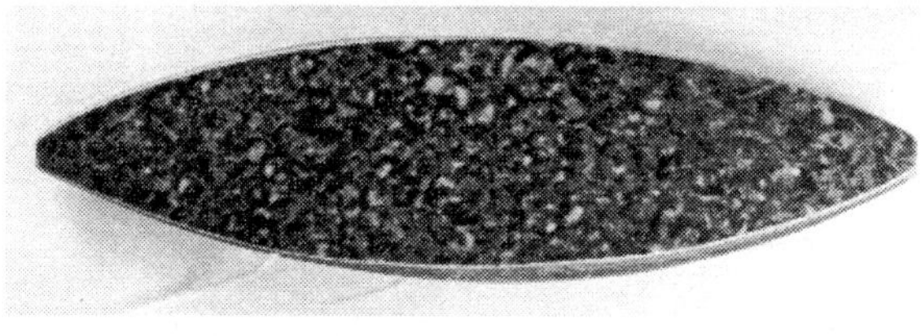

6-33 Top: 3 inch formica shuttle. Right: Corian mini-shuttle pendant, author's collection.

Chapter Seven

Wood Tatting Shuttles

The majority of antique American wooden tatting shuttles found today were hand crafted by the loving hands of folk artists. Often they were made by the tatters themselves out of necessity from native wood in their surrounding area. The sizes and types vary greatly with a range in quality from very basic to skillfully crafted. Unless dates were carved on the shuttles or they were documented by family members, their age cannot be determined.

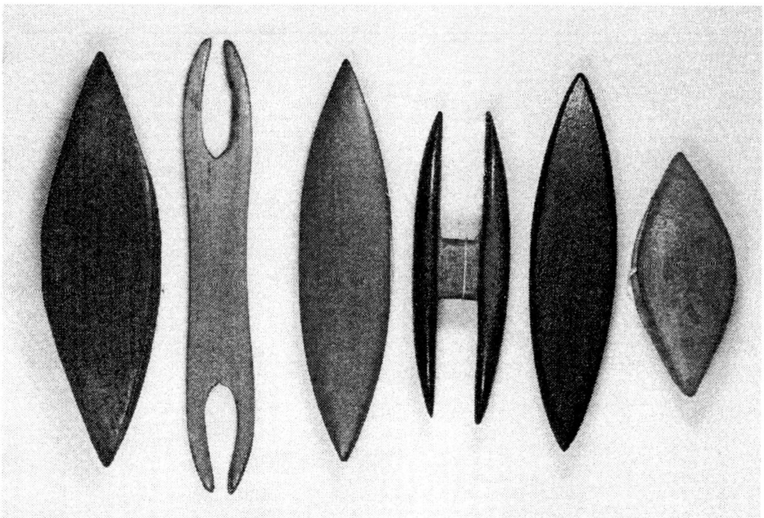

7-1 A variety of hand crafted wooden shuttles, Middleton collection.

7-2 Handmade shuttle with hook, Middleton collection.

7-3 Handmade shuttle, author's collection.

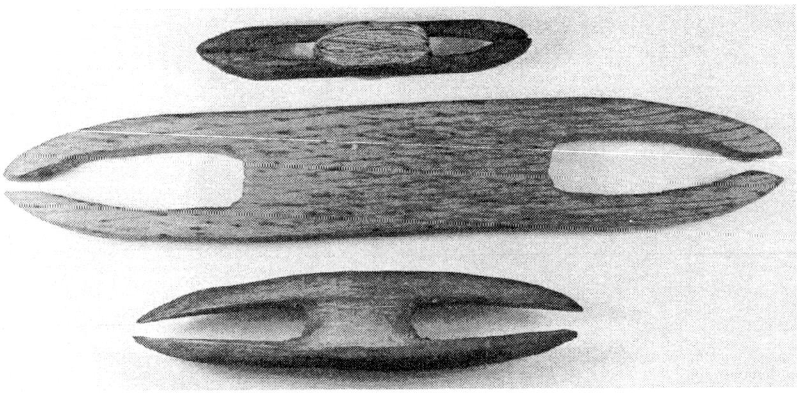

7-4 Hand crafted wood shuttles, Tollinger collection.

The oldest wooden shuttles that can be approximately dated are not American but rather Scottish. They were made from sycamore wood and sold as souvenirs beginning in the mid-nineteenth century. There were several different Scottish companies producing small wooden items, as well as tatting shuttles. Because the largest and most notable manufacturer was the Smith family of Mauchline, Scotland, Mauchline ware has become the generic term for these items.

There were three main finishes of Mauchline ware products, tartan ware, transfer ware, and fern ware. When the Smiths began making tartan ware tatting shuttles, a tartan pattern was drawn on paper then glued to the wood. They were stamped with the tartan name in gold lettering and painted with a numerous coats of varnish. Not long after tartan ware appeared, transfer ware became popular. Engravings and photographs depicting different locations in Scotland were soon followed by scenes of tourist sites in countries all over the world including America. Scenic transfer ware was most popular between 1860 and 1900, but was made until 1933.

7-5 Tartan ware shuttle, Tollinger collection.

7-6 Mauchline shuttle with Melrose Abbey transfer, author's collection.

Between 1870 and the late 1920s the Smiths of Mauchline developed fern ware. Originally they applied real ferns and later they used transfer pictures and photographic labels. Seaweed and colored ferns were also used as decorations. Another popular process was one in which ferns were dusted with a dark talc and removed to form a reverse stencil effect.

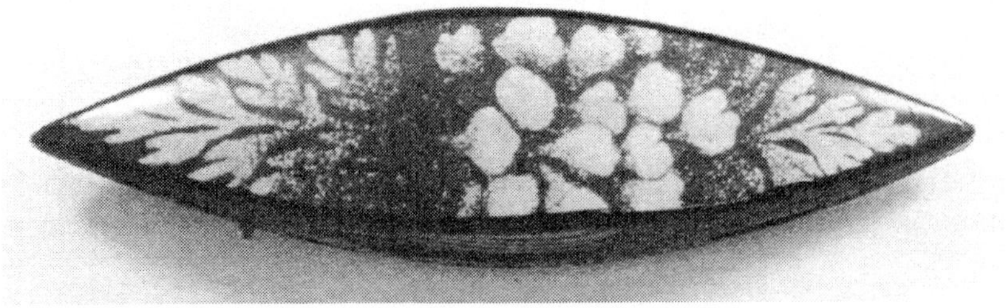

7-7 Fern ware shuttle, Demmer collection.

Many other wooden shuttles were imported to America. "Shuttle Art – The Magazine for Tatters," a small newsletter from the 1950s edited by Dorothy Tooker of West Hanford, Conn. advertised shuttles made of olive wood from Bethlehem, Jordan. There were also wooden shuttles from Denmark in Rosewood, Boxwood, Guaicumwood and Laburnum wood.

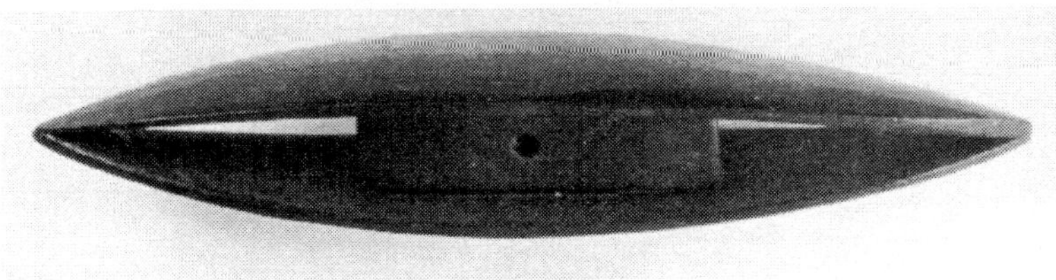

7-8 Wood shuttle, author's collection.

In 1972 Carol Winandy began selling wood shuttles. They were essentially 4 inch test models for her now popular Tatsy shuttle. They were made by her brother and sold through ads in the *International Old Lacers Inc. Bulletin* and the Serendipity Shop in *Workbasket.* The most recognizable of these are the 1976 U.S. Bicentennial commemoratives marked with the dates "1776 – 1976" and an American eagle symbol in a burnt wood technique. For more about the shuttles Carol Winandy created please refer to the chapter on Plastic Shuttles.

7-9 Facsimile of bicentennial shuttle by Carol Winandy.

Kermit Molde first developed a polished lilac wood jewelry line in the late 1960s. Around 1975 after a request by a Des Moines Tatting Guild member he began making shuttles. The lilac wood is cut locally and delivered to Mr. Molde by friends. Lilac is a very hard, dense wood, tan in color with the grain of the wood changing from a dusty rose to a brilliant burgundy color with age. Lilac's small branches take 2 to 3 years to dry naturally. Mr. Molde obtains a high degree of polish without varnish by using pumice stone and beeswax rubbed in with soft leather. The lilac shuttles are 2¼ inches in length without a point & 2 ½ inches with a point. The points on these shuttles are fiercely sharp.

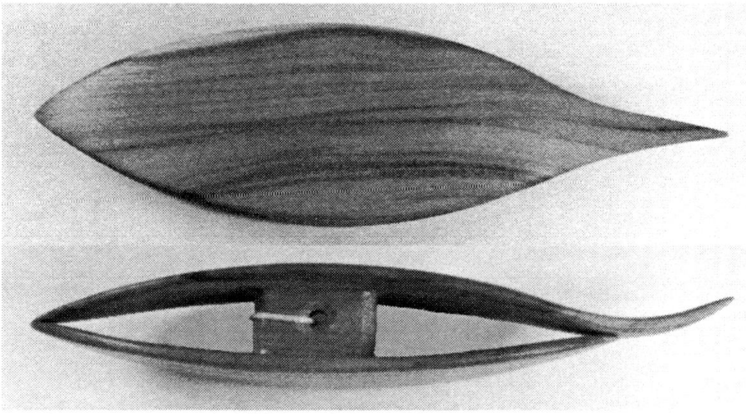

7-10 Lilac wood shuttles by Kermit Molde, author's collection.

Diana Andra is the exclusive shuttle maker at "Turn of the Century," a two member family workshop in Mansfield, Ohio. She designed a functional tatting shuttle suitable for production and sale in 1982. Ms. Andra uses a variety of exotic woods to make two sizes of shuttles with or without a hook, each fashioned from a single piece.

7-11 Shuttles by Diana Andra, author's collection.

When Elaine Stillwell, a self-taught tatter, could no longer find plastic shuttles she asked her husband to make her one of wood. Bluebird Woodcrafts incorporated early in 1982 after Leonard Stillwell developed a technique for the production of hand crafted shuttles. Mr. Stillwell made shuttles out of many different types of exotic woods, in 3 and 4 ½ inches lengths, with or with out a pick. Leonard Stillwell made hundreds of shuttles in his lifetime.

7-12 Bluebird shuttle by Leonard Stillwell, author's collection courtesy of Elaine Stillwell.

Shuttle Makers of the 1990s

Dennis Hand produced shuttles from exotic woods for a few years in the mid-1990s. The sides were steam bent then assembled with the center post and wooden point. The outside of these 3 inch shuttles were lightly varnished and the insides were not.

7-13 Hawaiian Koa wood shuttle by Dennis Hand, author's collection.

Richard Sietz passed on his knowledge of the bent wood process of shuttle making to his wife Georgia, after being shown by her step-grandfather, Al Goodwin. Their first wood shuttles, offered for sale in the winter of 1994, were of maple, mahogany, walnut and cherry woods, in two sizes, 2 ¾ inches and 4 ½ inches. At the 1995 I.O.L.I. convention their first white painted shuttle decorated with decals were sold. Over the next few years, different colors, specialty papers and fabric were added as decoration. In 1999 they offered shuttles made of exotic woods which are distinguished by brass pins in the center post. In 2000 the Sietz family began making flat shuttles from oak in two sizes. The elliptical shape came first, followed by fish-shaped shuttles that were offered in a variety of woods.

7-14 Wood shuttles by Georgia Sietz, author's collection.

Randy Houtz began making shuttles in 1993 when his brother Gary, an avid tatter, broke his last shuttle in the middle of a project while visiting. Encouraged by Gary to make more, the GR-8 shuttle construction began. The Houtz shuttles are approximately 2 ½ inches in length and about ¾ of an inch in width. They are individually hand crafted from a single piece of wood, deer, elk or moose antler, cow horn, abalone shell or ivory nut. His design is one that allows thread to be pulled from a bobbin held in the shuttle with a binding screw. The screw allows the tatter to adjust the tension on the bobbin. These shuttles are also made with two bobbins inside. Mr. Houtz has adorned some of his shuttles with inlays of charms and precious stones. They may also be found with chip carving and scrimshaw designs. Mr. Houtz shuttles are marked with an "8" on the outside of one end.

7-15 Grape wood shuttle by Randy Houtz, author's collection.

David Smith began hand crafting shuttles around 1995 after talking to tatters on an internet forum. Today he makes most of his shuttles one at a time by custom order because of the large number of choices he offers. Mr. Smith creates shuttles in a wide variety of exotic woods. There are five basic sizes ranging from 2 inches to 3 3/8 inches with no point. One may then choose to have either a wooden point, brass point or brass hook attached.

7-16 Walnut wood shuttle wtih brass point by David Smith, author's collection.

Chris Hinton started creating unique wooden shuttles in 2000. The open work of the sides are cut with a piercing saw, then steam bent and finely finished. He laser cuts a variety of designs in the center of one side. The opposite side has the initials TSS for "The Shuttle Shop" along with the edition number. The shuttles are currently offered in two sizes, 2 1/4 inches or 3 inches, with or without a point. He also specializes in inlays of contrasting woods, precious stones, abalone and gold.

7-17 Shuttles by Chris Hinton, author's collection.

Chapter Eight

Tatting Tools and Kits

Following the popular tatting art form were numerous inventions of tatting implements. The most useful tools to the tatter were those that made the monotonous job of winding the shuttle easier. There were eighteen U.S. Patents for shuttle winders and three U.S. Patents for shuttle holders. Listed below are photographs and illustrations of surviving patented tools.

Holders and Winders

The Ideal

The most widely advertised shuttle holder was the "Ideal." Patented by Percy Linderman in 1916, it was shown in a varicty of ladies' magazines as early as 1915. Later it was sold in combination with other items such as pattern books, shuttles or thread. The shuttle ends were slipped into the holder and when squeezed, one side of the shuttle

8-1 1916 Collingbourne advertisement.

would open to allow easy winding. This shuttle holder is being reproduced today by Lacis of Berkeley, California.

8-2 Ideal shuttle holder, author's collection.

Webster Company

The sterling silver shuttle holder by the Webster Co. was illustrated in catalogues of 1916 & 1917. The shuttle holders flat plate was designed with clips to hold the sides of a shuttle open so it could be wound with thread easily.

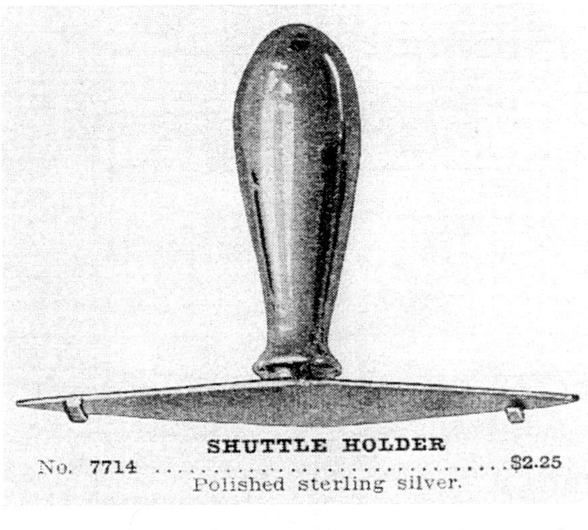

8-3 Webster Co. shuttle holder, 1916 G. W. Huntley catalogue, author's collection.

H. P. Jones Winder

The earliest shuttle winder known to this author is the H. P. Jones patent of 1867. The winders are a beautifully crafted of wood, with sliding rings on the holder that adapt to different sized shuttles. They are marked in the center with the words "Pat. June 1867."

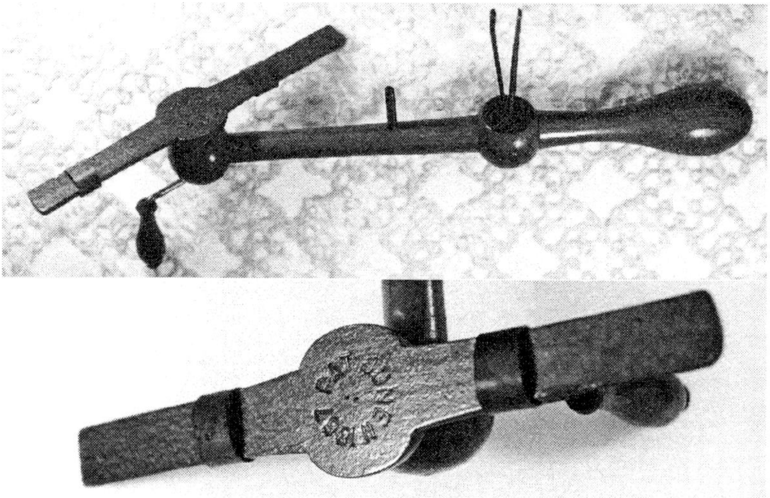

8-4 H. P. Jones patented shuttle winder, Tollinger collection.

J. B. Shuttle Winder

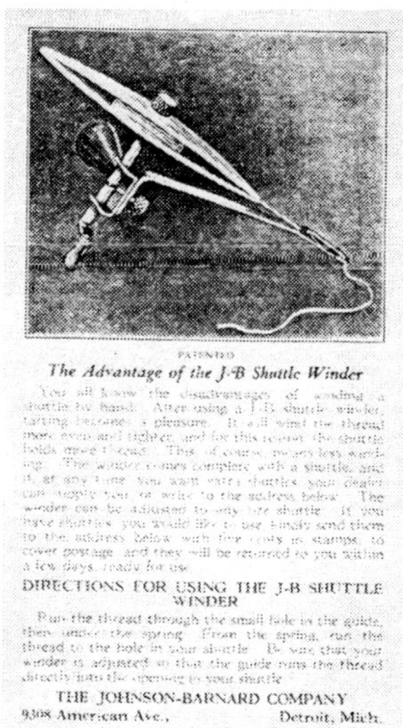

Daniel W. Johnson's patented winder of 1918 was manufactured by the Johnson-Barnard Company of Detroit, Michigan. Advertised as the J-B Shuttle Winder, it came with its own shuttle. The company's advertisement encouraged tatters to mail in their shuttles with the price of postage to have them drilled with center holes to fit the J.B. winder. Common sense would assume this did not happen too often. The winder was crafted of metal parts with wood handles.

8-5 J. B. shuttle winder, Woodyard collection.

The Gem Winder

The Gem shuttle winder may be seen in *Modern Priscilla* magazines of 1917. Wesley C. Jones of Portland, Oregon was granted the U. S. patent in 1918. The winder was sold in F. W. Woolworth stores and various mail order houses well before its patent date.

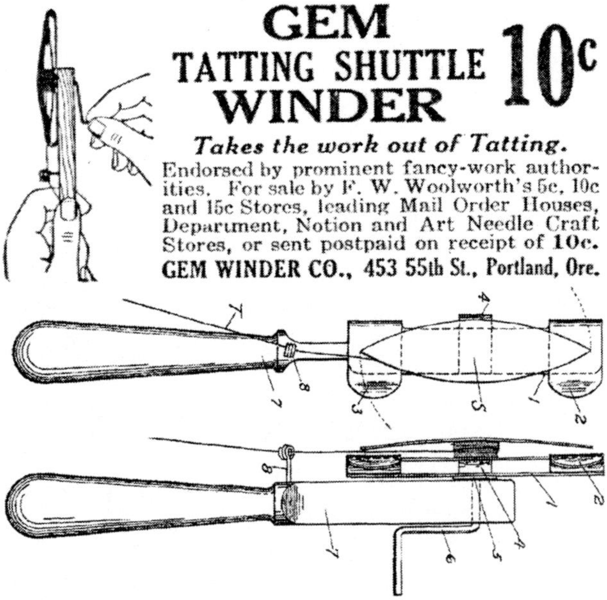

8-6 Top: Gem winder ad from 1917 *Modern Priscilla* magazine. Bottom: W. C. Jones' 1918 patent.

Mitchell Winder

Marshall Field & Company of Chicago, Illinois was selling John F. Mitchell's 1923 shuttle winder before its patent date. This wonderfully engineered metal shuttle winder bears the words "Pat. Applied For" across its main arm.

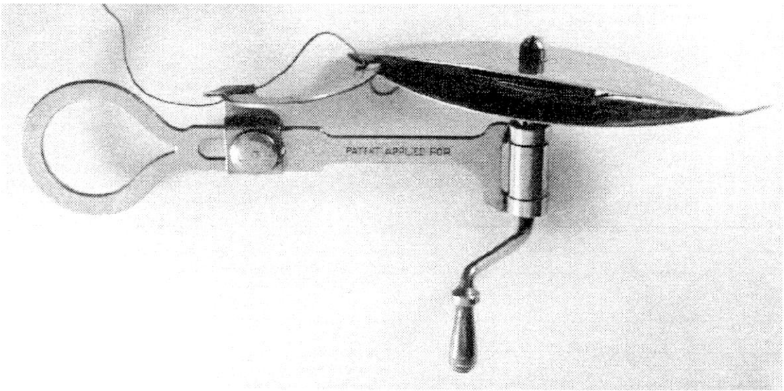

8-7 Mitchell patented shuttle winder, Tollinger collection.

C. D. Fischer Winder

The shuttle winder by Charles D. Fischer of Denver, Colorado was patented in 1924 and sold with its own shuttle. It was widely advertised in ladies' magazines. More examples of the shuttle have been found than of the winder. Both the shuttle and winder are clearly marked with the C. D. Fischer name. The C. D. Fischer Company went out of business in 1932.

8-8 1926 *Needlecraft* advertisement.

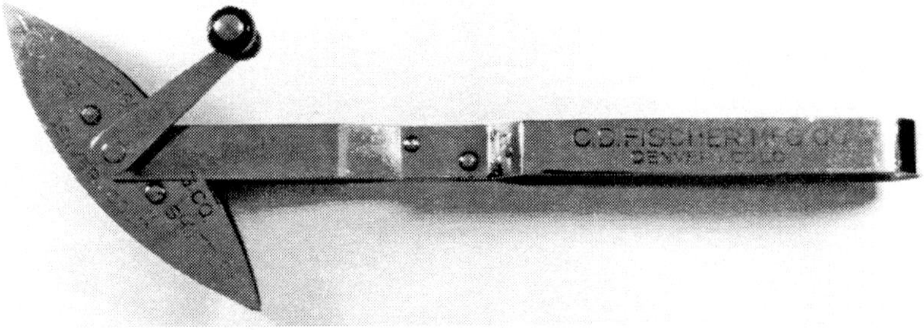

8-9 C. D. Fischer shuttle and winder, Middleton collection.

Joining Tools

When there is no point on the end of a shuttle, the tatter needs a pick of some sort to join the lace. In Europe most shuttles did not have a point and in tatting boxes one would find either a small crochet hook or a ring and chain with a small stiletto or crochet hook attached. Small crochet hooks are often found in the workboxes of American tatters as well.

8-10 Small crochet hook used to join tatting, author's collection.

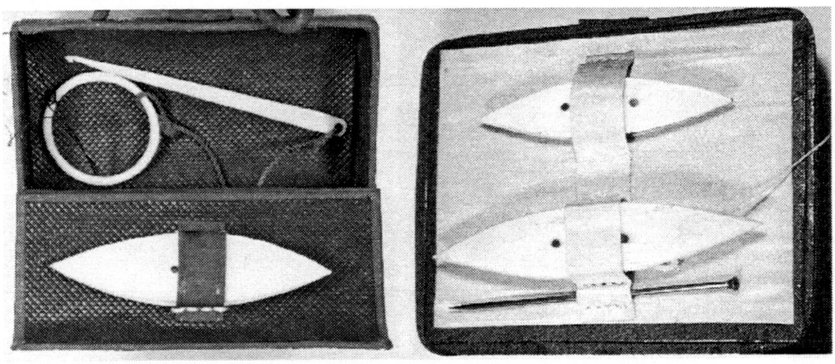

8-11 English tatting boxes, Middleton collection.

Inventions were also patented to replace simple joining tools. Figure *8-13* shows the 1917 Charles Opperman patented tatting ring with a hook attached. There were also three patents with hooks attached to thimbles as tatting tools. No examples of these have been found to date. When a hook wasn't handy, tatters often used a simple hairpin, similar to the one in *fig. 8-15* designed for keeping pince-nez glasses from dropping.

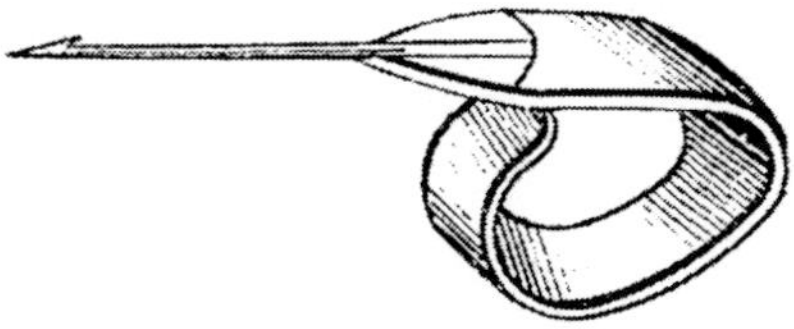

8-12 1917 Opperman patent for a tatting implement.

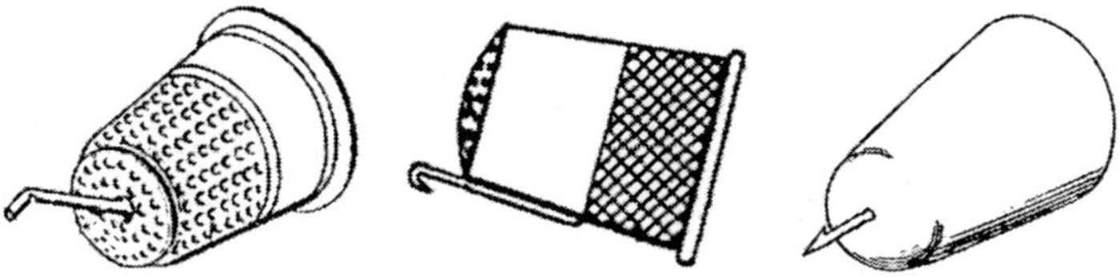

8-13 Patents for tatting thimbles, Left to right 1918 E. M. Gourley, 1920 L. L. Bauman & 1921 M. C. Mouser.

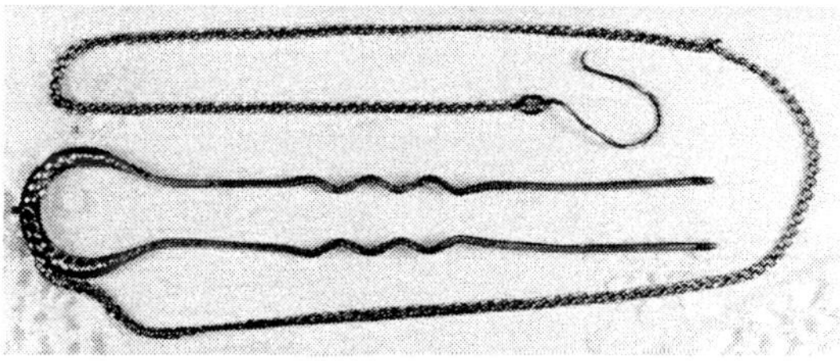

8-14 Left: hairpin on pince-nez chain, Right: woman wearing pince-nez glasses with hairpin, author's collection.

Rings

Tensioning rings designed to be worn on a finger were invented to help keep the thread from slipping while tatting. Examples of the two advertised in *Modern Priscilla* magazines in 1916 are shown in *fig. 8-17* Harry Samuels of Des Moines, Iowa received a patent for his tatting

8-15 "Evenit" tensioning ring. author's collection.

tensioning ring in 1916. A similar ring was sold in *Workbasket* magazine in the 1950s called the "Evenit." This ring was manufactured by Crane Products and sold by Plastic Container Inc. of New York.

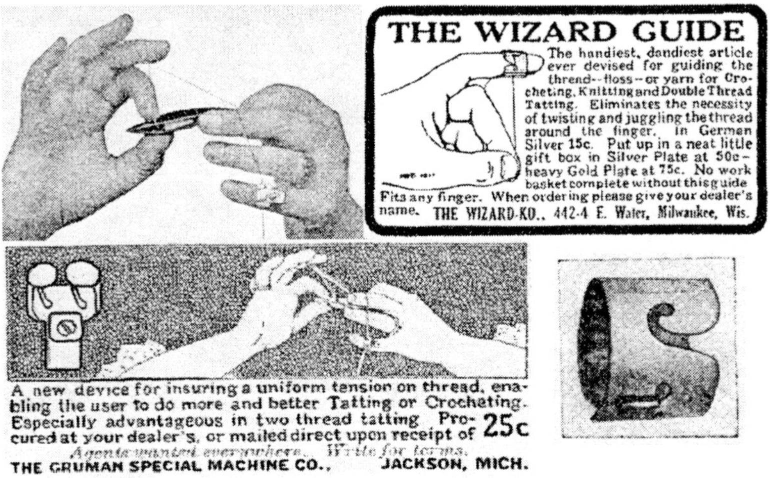

8-16 1916 & 1917 *Modern Priscilla* advertisements.

Thread Holders

Other tatting implements that may be found were containers designed to hold a ball of thread while tatting or winding the shuttle. The J. P. Coats' "Tat-It" ball is often found. It is two inches in diameter, made of clear plastic with the words "J. P. Coats' Tat-It" embossed on the top. The sides separate to insert the thread and a small hole in the top dispenses the thread. The "Tat-It" was sold with boxes of tatting thread from 1920-1940.

8-17 Coats & Clark "Tat-It" ball holder.

Other types of thread holders were wristlet varieties. These were made of metal, celluloid or both. They had a bracelet which slipped over the wrist and usually a rod with a stopper to hold the thread while the tatter was working.

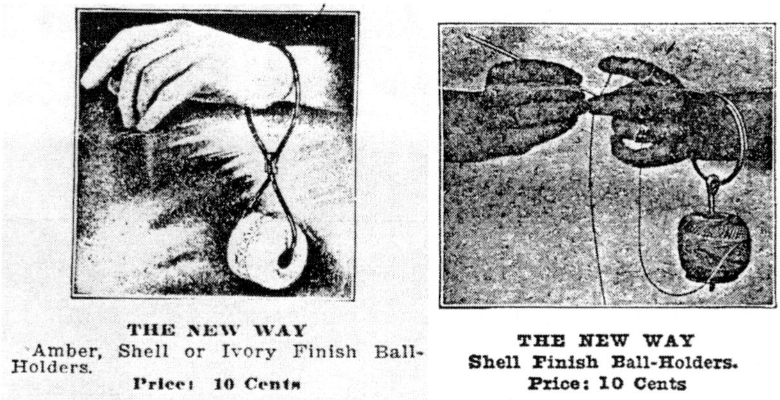

8-18 Ball holders advertised in Collingbourne's magazine 1915 & 1916.

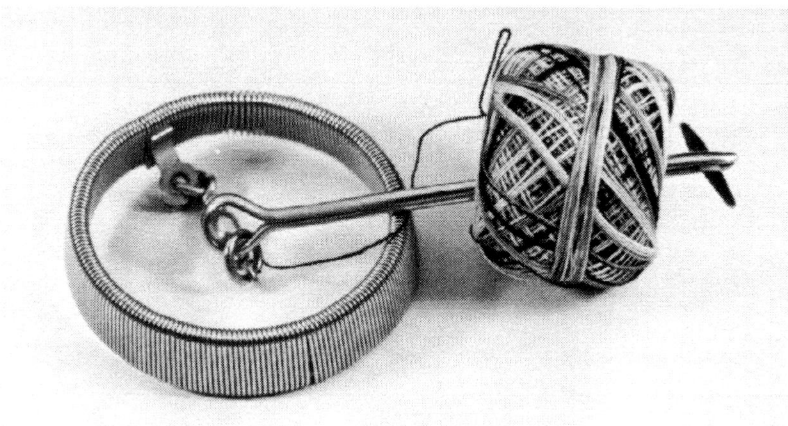

8-19 Ball holder marked "U.S. Pat Appl. For.," author's collection.

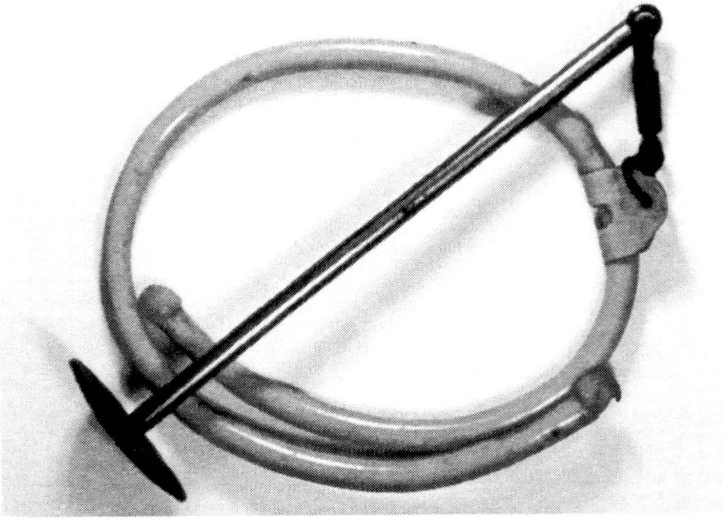

8-20 Celluloid and brass wristlet ball holder, author's collection.

Sewing Kits

Tatting shuttles were included in many sewing kits from 1915-1940. Some kits were small with only a shuttle and small crochet hook. Others may have included a bodkin and stiletto and still others might have had a whole range of sewing tools with needles, thread, holders, scissors and crochet hooks. Shuttles were usually bone, sterling or celluloid. One of the major sewing kit makers was C. J. Bates and Sons. They were also well known for their manicure sets. Their kits had a small tag with the name C. J. Bates and Sons sewn into the lining.

8-21 Left & Right: 1923 Levin Bros. Catalog, courtesy of UCSB, Romaine collection, Center: 1917 Baird & North catalog, courtesy National Museum of American History, Warsaw collection.

8-22 C. J. Bates company sewing kit, Oster collection.

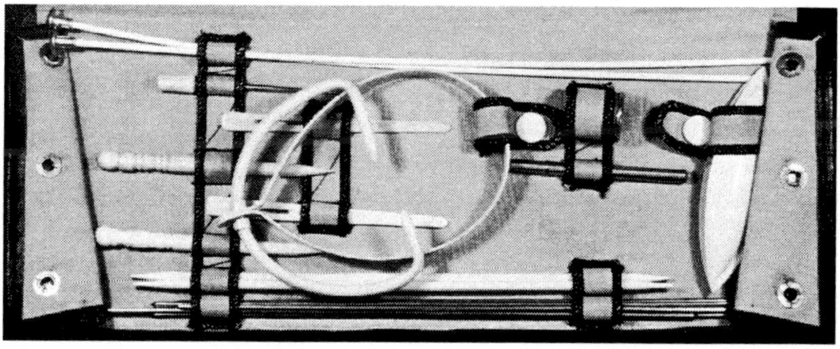

8-23 Sewing kit with shuttle and thread holder, author's collection.

Most Wanted List

The following graphics show the patented shuttle winders that have yet to be found and may not have been made. They are shown for those collectors interested in searching for the hard to find.

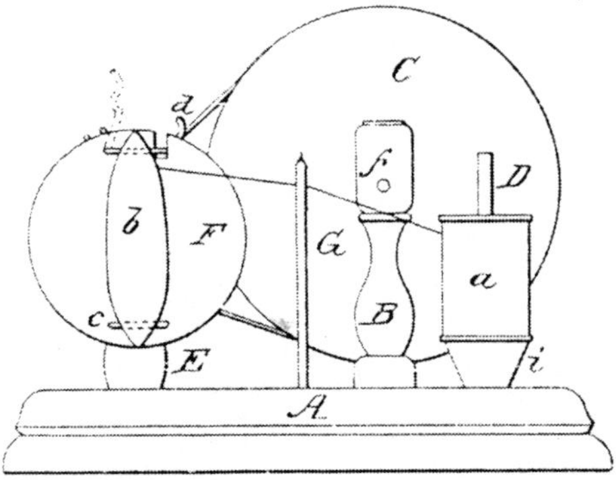

1868 U.S. Pat. #83,921

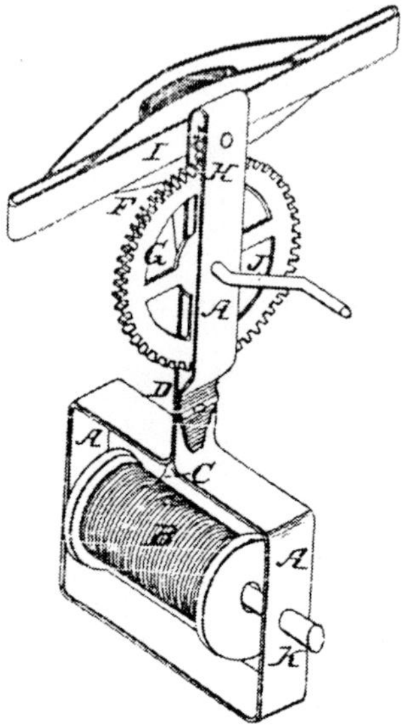

1866 U.S. Pat. #60,407

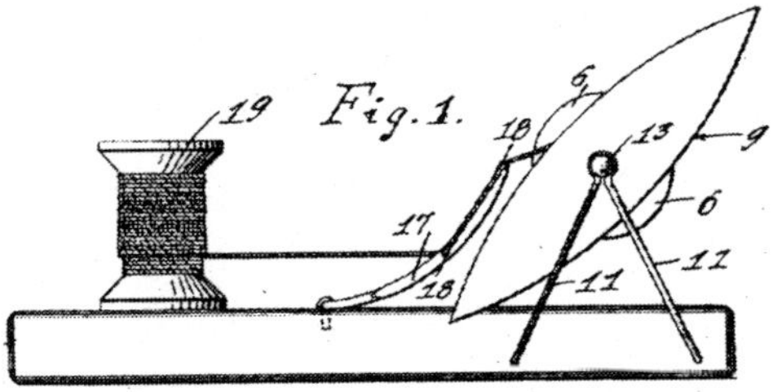

1915 U.S. Pat. #1,161,624

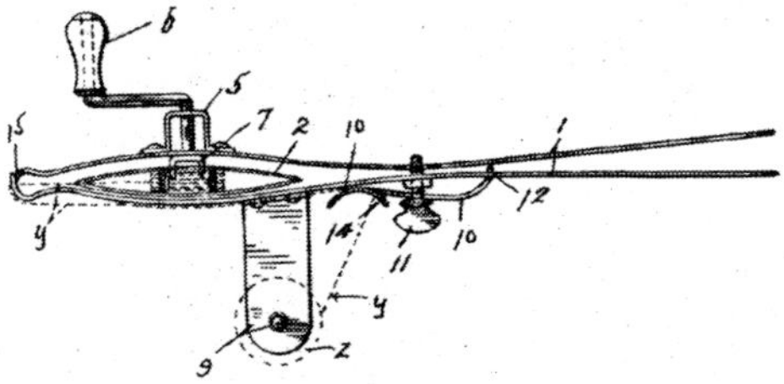
1916 U.S. Pat. #1,170,586

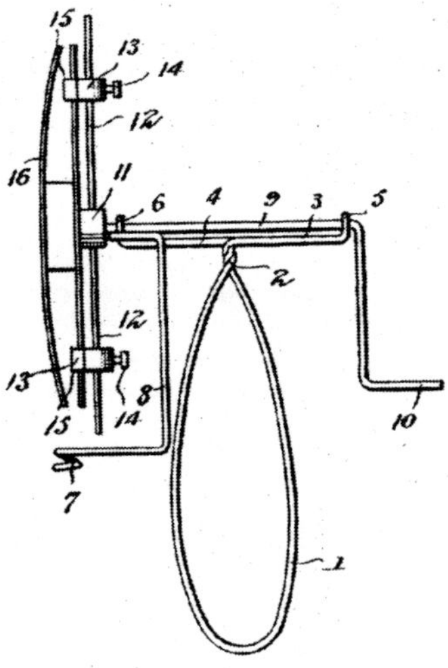
1916 U.S. Pat. #1,178,321

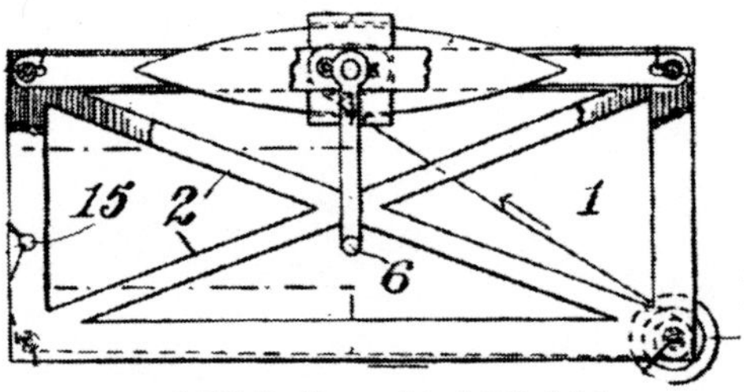

1916 U.S. Pat. #1,195,501

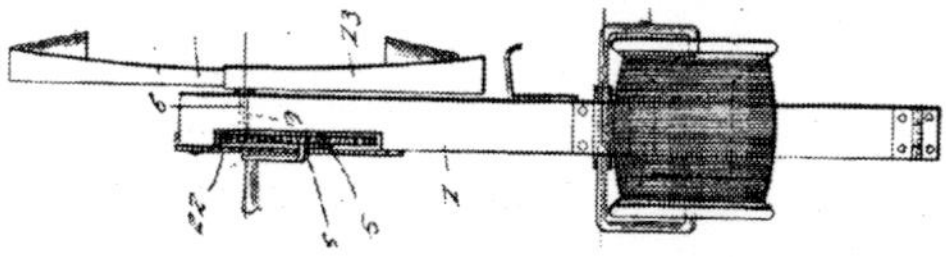

1917 U.S. Pat. #1,216,994

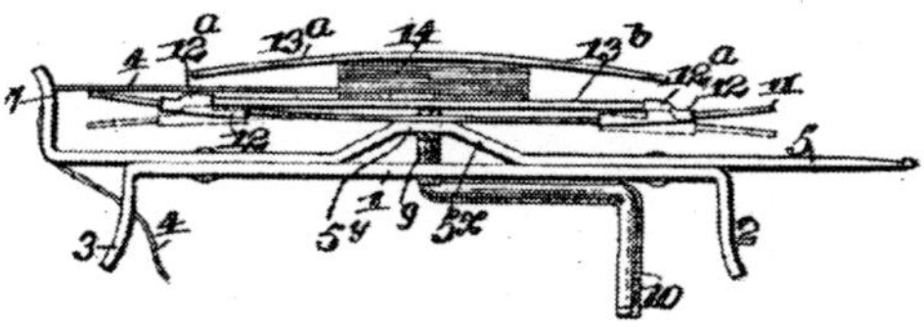

1917 U.S. Pat. #1,248,243

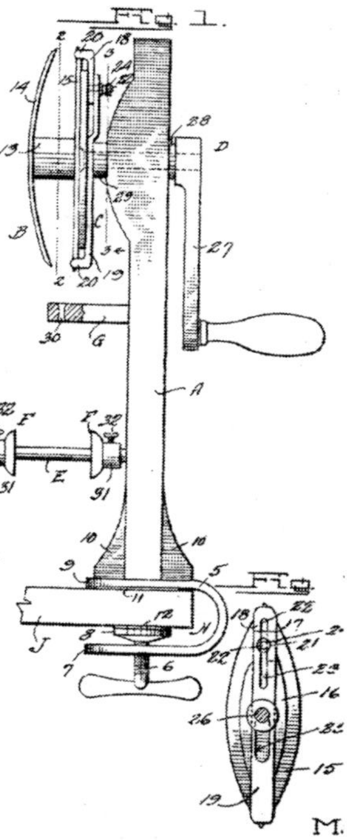

1917 U. S. Pat. #1,246,586

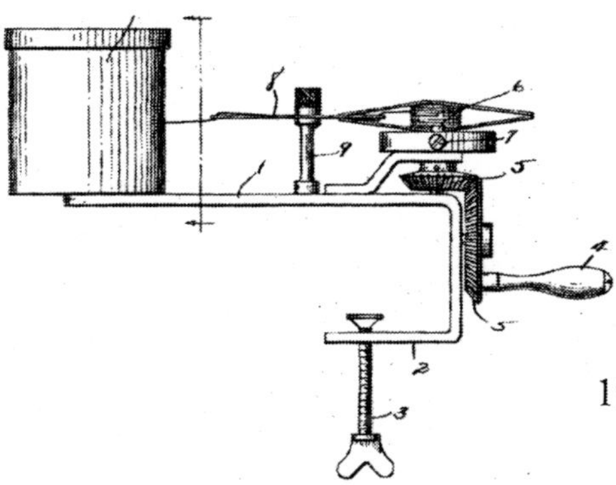

1921 U.S. Pat. #1,448,619

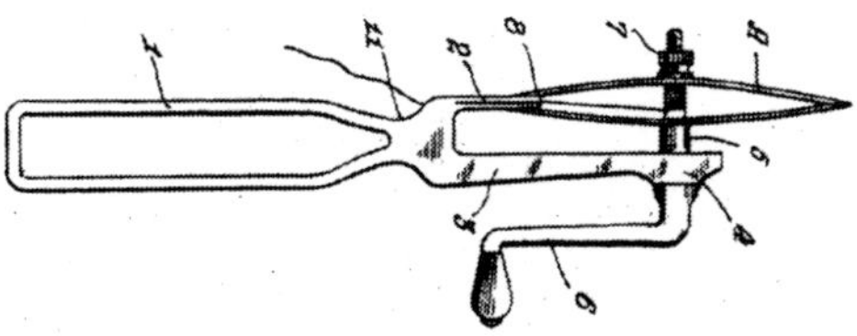

1925 U.S. Pat. #1,528,742

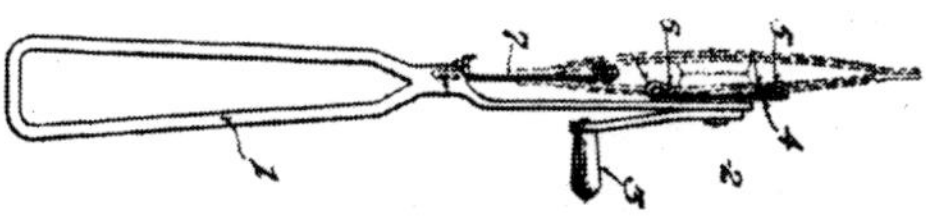

1926 U.S. Pat. #1,569,367

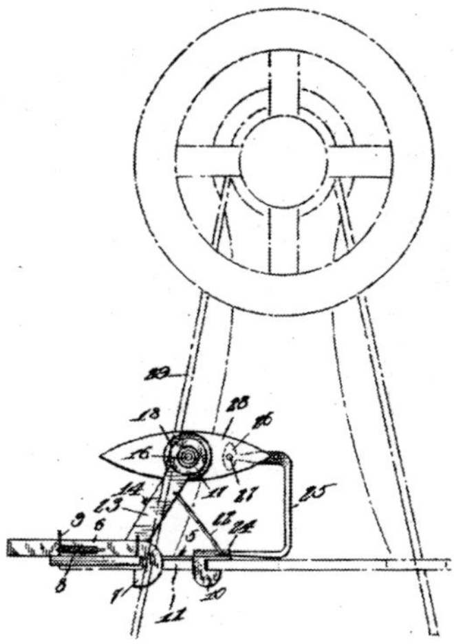

1924 U.S. Pat. 1,492,032

Chapter Nine

Collecting Sense

1. Know your own collection. There comes a point in time when so many tatting shuttles have been collected that remembering the particular colors and sizes is impossible. A solution for this may be to reproduce them on a color photocopier. Carefully put as many shuttles on the copier as possible and cover with either a 8 by 11 ½ or 14 inch paper. Photocopies not only provide a reminder of the colors but also the sizes and may be taken along on shuttle searches. A photocopy is also a good way to record shuttle purchase information.

2. Ask permission before handling another collector's or dealer's objects and always be very careful while examining them.

3. When shopping for shuttles, one should carry either a magnifying glass or a jeweler's loupe. This is particularly helpful when trying to read hallmarks and inspecting for irregularities.

4. Examine the shuttles in good light and always inspect for damage. As with most antiques and collectibles the condition of an object affects its worth.

5. Be wary of the following:

a. Shuttle ends not together. Do not buy a shuttle that does not have its ends closed unless it is of sterling, gold or one that is hard to find. When one buys a shuttle in poor condition, it should always be with the intention of replacing it eventually with another in better condition. Sterling and gold shuttles are made of softer metals that can be gently bent back into shape.

b. Chips and cracks. Especially in bone, ivory, shell and horn shuttles, cracks can become larger in time and possibly even break.

c. Discoloration. This may occur from price stickers, heat and liquids. Depending on the degree of damage, ask for a discount or don't buy at all.

d. Crazing. This is serious damage to celluloid shuttles and is seen mostly on center posts. Crazing is a deterioration of celluloid which means it inevitably crumbles and turns to dust. Sometimes crazing appears as little creases in the middle of the sides of a shuttle. Never buy a shuttle with this defect.

e. Minor scratches and dents. These should be considered less important. Small dents on the ends of shuttles with no point are usually caused by a tatter who held the shuttle in her teeth while joining the tatting. These dents will be charming to some and distracting to others.

6. Dealers statements. Even dealers with good intentions can often be wrong about what a shuttle is made of or when it was made. Don't be afraid to question any statement made by a dealer.

Bibliography

Andere, Mary, *Old Needlework Boxes and Tools*, Great Britain, David & Charles, 1971.

Attenborough, Bessie M., *The Craft of Tatting*, Great Britain, Bell & Hyman Lmt., 1979.

Baker, John, *Mauchline Ware*, Shire Publications Ltd., England 1985.

Baker, Lillian, *Twentieth Century Fashionable Plastic Jewelry*, Collector Books, Paducah, KN, 1992.

Barker, P. W., *Rubber Industry of the United States 1839-1939*, Washington DC, United States Government Printing Office, 1939.

Burack, Benjamin, *Ivory and Its Uses*, Rutland, Vermont, Charles E Tuttle Company 1984.

Doussy, Michel, *Antiques, Professional Secrets for the Amateur*, New York: Quadrangle/The New York Times Book Co., 1971.

Dubois, J. Harry, *Plastics History U.S.A.*, Cahners Books, Boston, MA, 1972.

Friedel, Robert, *Pioneer Plastic, The Making and Selling of Celluloid*, The University of Wisconsin Press, Madison, Wisconsin, 1983.

Groves, Sylvia, *Needlework Tools and Accessories*, New York, NY, Arco Publishing, 1966.

Haynes, Williams, *Cellulose, The Chemical That Grows*, Doubleday & Company, Inc., Garden City, NY, 1953.

Hughes, Elizabeth & Lester, Marion, *The Big Book of Buttons*, Maine: New Leaf Publishers, 1991.

Johnson, Hugh S., *The Blue Eagle From Egg to Earth*, Garden City, NY, Doubleday, Doran & Company, Inc., 1935.

Leominster City Directories, courtesy of Leominster Public Library, Historical & Genealogy Collections.

Little, Nina Fletcher, *Neat and Tidy Boxes and Their Contents Used in Early American Households*, New York: E.P. Dutton, 1980.

McManus, Michael, *A Treasury of American Scrimshaw*, New York: Penguin Books USA Inc., 1997.

—, *New Brunswick and its Industries*, New Brunswick, NJ, A. E. Gordon Times Printing House, 1873.

Pinto, Edward & Eva, *Tunbridge and Scottish Souvenir*, London, G. Bell & Sons, 1970.

Penniman, T. K., *Pictures of Ivory and other Animal Teeth, Bone and Antler*, Oxford, England, Oxford University Press, 1952.

Rainwater, Dorothy T., *Encyclopedia of American Silver Manufacturers*, New York: Crown Pubs., Inc., 1975.

Romaine, Lawrence, *A Guide to American Trade Catalogs 1744-1900*. New York: R.R. Bowker Co., 1960.

Secretary of State, Compiled by the, *Corporations of New Jersey, List of Certificates to December 31, 1911*, Trent, NJ, MacCrellish & Quigley, 1914.

Schidrowitz, P. & Dawson, T. R., Editors, *History of the Rubber Industry*, Compiled under the auspices of the Institution of the Rubber Industry, Cambridge, England, W. Heffer & Sons Ltd., 1952.

Stage, Sarah, *Female Complaints*, New York: W.W. Norton & Company, 1979.

Strohmeyer, Alice, *Tatting Shuttles*, date unknown.

—, *Thomas Register of Manufacturers*, 1905-2000.

Time-Life Books, *The Encyclopedia of Collectibles*, 1987.

Whiting, Gertrude, *Tools and Toys of Stitchery*, New York, NY, Columbia University Press, 1928.

Woodyard, Sandy, *Tatting Shuttles*, date unknown.

Zalkin, Estelle, *Zalkin's Handbook of Thimbles and Sewing Implements*, Radnor, Penn., Warman Publishing Co., Inc., 1988.

Catalogues and Periodicals

American Wholesaler Corp. September 1925.

Baird-North Co. Catalogues, 1906-1928.

Bellas Hess & Co. Catalogue, New York,. 1921-22.

Bucilla Blue Book of Novelty & Filet Crochet, Bernhart Ulmann Co. Inc., New York, 1916.

Bucilla Catalog of Everything for Art Needlework, Fall and Winter, 1940.

Butler Bros. Catalogue, New York, 1920-1939.

Charles Burroughs Co., Bakelite Co. Exhibition, 1920, Du Bois Collection.

Charles Williams Catalogue, 1914-1923.

Collingbourne's Encyclopedia of Technologic Art Needle Work Instruction, Elgin, IL, 1915 & 1916.

Daniel Low & Co. Catalogues, Salem, Mass. 1898 –1929 Courtesy of the Peabody Essex Museum, Salem Massachusetts.

F. Herrschner Catalogues, Chicago, IL, 1913 -1993.

Geo. T. Brodnax, Inc., Memphis, Tenn., 1916.

Goodyear Rubber Co. Trade Journal, New York, 1878.

Gray & Dudley Co., 1916.

Jason Weiler & Sons Catalogue, Boston, MA, 1927.

John M. Smyth Mdse. Company Catalog #70, Chicago, IL, 1914.

Jones Quarterly Catalogue, Spring 1881, New York.

Kresge's Katalog of the World's Best Merchandise, 1913-1915.

Ladies Home Journal, 1915.

McCall's Needlework, Spring-Summer 1951.

Modern Priscilla magazine, 1911-1916.

Montgomery Ward & Co. Catalogues, 1895-1932.

N. Shure Co. Catalog, Chicago, IL, 1918.

Needlecraft magazine, 1915-1923.

Otto Stumpf & Co., Chicago, 1911.

Pohlson Gifts Catalogue, Pawtucket, Rhode Island, no date.

Richardson's Tatting Book No. 7, undated.

Sears, Roebuck and Co., 1902-1946.

Spring and Summer Catalogue, John Wanamaker, 1905.

Successful Merchandising General Catalog of 1927, Watts, Ritter & Company, Huntington, WV.

T. Buettner & Co. Inc. Catalog, Chicago, IL, Fall 1939.

Webster Catalogue, 1913 & 1914.

Williams-Richardson Co., Ltd., New Orleans, LA, 1925.

Your Bargain Book, Charles William Stores, New York City, 1916, 1917.

Index

H

I

J

K

L

M

N

P

R

S

T

U

V

W